RL = 7.9
4 points

ILLUMINATED PROGRESS

the story of
Thomas Edison

Illuminated
Progress
the story of
Thomas Edison

Roberta Baxter

MORGAN
REYNOLDS

PUBLISHING

Greensboro, North Carolina

the *Profiles*
IN SCIENCE

series includes biographies about . . .

Nikola Tesla   Isaac Newton
Louis Pasteur   Robert Boyle
Mary Anning   Rosalind Franklin
Lise Meitner   Ibn al-Haytham
Tycho Brahe   Edmond Halley
Johannes Kepler   Marie Curie
Nicholas Copernicus   Caroline Herschel
Galileo Galilei   Thomas Edison

ILLUMINATED PROGRESS
THE STORY OF THOMAS EDISON

Copyright © 2009 By Roberta Baxter

Library of Congress Cataloging-in-Publication Data

Baxter, Roberta.
    Illuminated progress : the story of Thomas Edison / by Roberta Baxter.
        p. cm. -- (Profiles in science)
    Includes bibliographical references and index.
    ISBN-13: 978-1-59935-085-1
    ISBN-10: 1-59935-085-8
    1. Edison, Thomas A. (Thomas Alva), 1847-1931. 2. Inventors--United States--Biography. 3. Electric engineers--United States--Biography. I. Title.
    TK140.E3B35 2008
    621.3092--dc22
    [B]
                                        2008007411

Printed in the United States of America
First Edition

Contents

Thomas Edison *(Library of Congress)*

one
Young
Businessman

O n New Year's Eve, 1879, trains pulled into the depot
at the tiny town of Menlo Park, New Jersey. Hundreds
of people disembarked the trains, and more came by
wagon and horseback. They gathered at the workshop and
laboratory of Thomas Alva Edison, and stared in amazement
at the sight before their eyes.

They saw white, steady light streaming from bulbs mounted
on poles along the path from the depot to the laboratory.
Nearby houses and a boardinghouse were also lit. Lights
shone in the yard of the laboratory and from the windows
of the building.

Inside the building, twenty-five more bulbs lit the laboratory tables, shelves of chemicals, and books in the library.
They illuminated the inventor himself, Thomas Edison.

During the evening, more visitors cascaded through the
laboratory, enjoying the demonstrations by Edison and his

team. They saw electric motors that ran a sewing machine and a water pump. One light bulb was stuck in a jar of water and burned for four hours. Lights were switched on and off to the amazement of people accustomed to gas lights that had to be lit and then blown out.

A few unexpected things occurred. One woman got too close to a generator and its magnetic field yanked the hairpins out of her hair. An employee of a gaslight company tried to sabotage the lighting system. But overall, the audience was thrilled by what they saw. The clear, steady light impressed them with its brilliance and its lack of fumes.

By the time the laboratory was closed, more than 3,000 people had celebrated the most unusual New Year's Eve. This demonstration of Edison's lights and electrical system was only the beginning.

Thomas Edison is known as the most prolific inventor of all time. His inventions stand at the beginning of three major industries, lighting our homes and entertaining us. He was famed throughout the world; people idolized and yearned to meet him. In today's world, he is still remembered as a man with an innovative mind and a capacity for hard work to make his dreams happen.

Sixteen years before Edison's birth, Michael Faraday discovered electromagnetism: the method to make electricity using magnetism. Ten years earlier, Samuel Morse had demonstrated his alphabet for the telegraph. Both would profoundly influence Thomas Edison. As he built on the work of others, inventors after him stood on his shoulders. Because of Edison, the world is a brighter, more productive place.

Before Thomas's birth, the Edison family had been on both sides of revolutions. During the American Revolution, the family had been Loyalists, siding with the British. Thomas's great-grandfather, John Edison (he spelled the name Edeson) was a wealthy landowner in New Jersey. He not only supported the British during the Revolution, but served as a scout for them. American rebels captured him and he was in danger of being hanged. Members of his wife's family were fighting in the Continental Army and they negotiated his release. After the war, his property was confiscated and the family fled to Canada to escape the condemnation of their neighbors.

In 1837, Samuel Edison participated in a revolt against the British-controlled government of Canada. A small group of men led by William MacKenzie tried to kidnap the governor of Toronto and set up their own government. Samuel Edison and others participated in the revolt in the western part of what is now Quebec. The revolt failed and Samuel fled Canada to escape arrest by royal troops. He settled in Milan, Ohio just south of Lake Erie. After a few months, his family joined him there. Edison later said of his father, "he has always been a rebel, a regular red-hot copperhead Democrat, and General Jackson was his hero."

On February 11, 1847, Thomas Alva Edison was born. He had a tiny body and a large head, so the doctor warned his parents that he might not live long. The doctor feared that the new baby had "brain fever" or meningitis. Thomas was the seventh child in the family, and the family already knew the tragedy of losing a child. Six-year-old Charlie had died in 1841. Samuel and Eliza both died before their third birthdays. When he was born, his sister Marion was eighteen, his brother Pitt was sixteen, and his sister, Harriet Ann (called Tannie) was fourteen. For most of his childhood,

Samuel Edison, Father of Thomas A Edison

Elizabeth Yukon Edison
Mother of Samuel Edison

Nancy Elliot Edison
Wife of Samuel Edison

The Edison Home at Milan, Ohio, Where on
February 11, 1847 Thomas A Edison Was Born

A collage of photos made in 1924 depicting Edison's childhood home,
mother, father and paternal grandmother.

(Courtesy of New York Public Library)

Edison lived as an only child because his siblings were so much older.

The new baby was named Thomas for an uncle and Alva for a friend of his father. In honor of the friend, he was called Alva or Al. As the months went on, the baby was healthy and growing.

Samuel Edison ran a small lumber mill and shingle factory. The town Milan was located on a canal that ran from the Huron River to Lake Erie. The huge amount of wheat loaded from wagons to barges every day made Milan one of the biggest wheat-shipping ports in the world. It was said that only Odessa, Russia, on the Black Sea shipped more wheat. However, the prosperity would not last.

Alva was a curious and adventurous child, in spite of frequent illnesses. His mother would punish him, but he would soon be in trouble again. He almost drowned in the canal once; another time, he almost suffocated in a grain elevator. When he was six, he started a fire in his father's barn. The flames quickly spread and the barn was destroyed. When questioned, Alva said he just wanted to see what the fire would do. Samuel marched him to the town square and whipped him in front of townspeople to stress the seriousness of Alva's offense.

Another event had even more serious consequences. Alva and a friend went swimming in a nearby creek. The other boy disappeared and as Alva later recalled, "I waited around for him to come up, but as it was getting dark I concluded to wait no longer and went home." Later when the boy's parents had been searching for him, Alva was awakened and he told what had happened. His friend had drowned and his body was later found. Most likely, Alva was too young to realize why the boy had disappeared, but even if he had told some-one right away the boy probably still would have died.

Always full of questions, Alva asked his mother why a goose was sitting on a nest of eggs. His mother explained that the goose was hatching the eggs. Later she found Alva in a neighbor's barn snuggled up to some goose and chicken eggs, trying to make them hatch.

As a young boy, Alva watched covered wagons drive through Milan, and his adventurous spirit caused him to imagine riding along with them all the way to the gold fields of California.

As railroads expanded across the country, trains became the preferred method for shipping grain. The canal boats carried less and less wheat. The prosperity of the town and Samuel's business declined. When Alva was seven years old, the family moved to Port Huron, Michigan. Samuel began his lumber business again and whatever other business venture captured his attention. When one enterprise wasn't success-ful, he just started something else. Most of his businesses earned little money, and the family was often in poverty because of the failures.

By the time of the move, Alva was the only child living at home. His oldest sisters had married and his brother had moved out on his own. Another sister had died, meaning that only four of the seven children survived.

Soon after Alva and his parents moved to Port Huron, Alva got scarlet fever. The disease damaged his hearing and left him vulnerable to respiratory infections for the rest of his life.

After recovering from the illness, Alva was enrolled in Reverend George Engle's private school. There were no public

Alva in 1857, around the time the family moved to Port Huron, and he lost part of his hearing to scarlet fever. *(Courtesy of M Stock/Alamy)*

schools in Port Huron at the time. In the prevalent method of the time, Reverend Engle expected his students to learn by memorizing facts. Questions were discouraged. If a student did not sit quietly, he was punished by whipping. This type of classroom was not a good situation for Alva. His curiosity led him to question everything, and his abundant energy made him squirm. When he was bored, he would gaze out the window and not pay attention.

Alva's parents decided to remove him from the school. Reverend Engle considered Alva a dull, uncooperative student, and the boy was learning little for the $130 tuition his parents paid.

Nancy Edison had been a teacher before her marriage, so she began to teach Alva at home. Although she also would whip Alva for misbehavior, she gave him lessons that suited his curiosity and energy. As was common for the time, much of Alva's lessons would come from the Bible. His mother also read to him from books by Charles Dickens, Shakespeare, and Edward Gibbon's *Decline and Fall of the Roman Empire,* and David Hume's *History of England.* He became a voracious reader on his own. "My mother taught me how to read good books," he later said, "and as this opened up a great world in literature, I have always been very thankful for this early training."

Nancy Edison was a highly religious woman. She attended church services regularly and took Alva along to Sunday school. Alva's father Samuel, on the other hand, was not religious. He believed much the same things as one of his heroes, Thomas Paine. During the Revolutionary War, Paine had published *Common Sense,* a pamphlet that helped inflame fervor in the colonies. Another book by Paine guided Samuel:

Thomas Paine *(Library of Congress)*

Age of Reason. As a freethinker, Paine opposed the hierarchy of religion and stated that churches existed only to terrify people into good behavior. He said, "The world is my country; to do good my religion." Alva, reading books from his father's library, was heavily influenced by this as well.

Alva later wrote,

> My father had a set of Tom Paine's books on the shelf at home. I must have opened the covers about the time I was thirteen. And I can still remember the flash of enlightenment which shone from his pages. It was a revelation, indeed, to encounter his views on political and religious matters, so different from the views of many people around us. Of course I did not understand him very well, but his sincerity and ardor made an impression upon me that nothing has ever served to lessen.

Alva also was inspired by Paine's words, "under all discouragements, [man] pursues his object and yields to nothing but impossibilities."

When Alva was eleven, a public school opened in Port Huron. He attended for a short time, probably to enhance his science education. But the teaching methods were the same as those of Reverend Engle. In a prank, Alva and another boy dropped a fishing line out of a second story window, hooked a chicken, and pulled it through the window. This incident, as well as several others, led to whippings by the principal. Alva recalled later that one day he heard a teacher call him "addled" and once he told his mother, he never returned to school.

One benefit from the public school was a book that Alva discovered: *A School Compendium of Natural and Experimental Philosophy*. The book included simple experiments and Alva began trying them all. At first, he set up a laboratory in his bedroom, but his parents requested he move it to the basement, hoping to alleviate some of the awful smells.

Samuel Edison continued to switch from job to job. When one didn't work out, he started something new. Alva learned resiliency from his father. One project they tackled together: Samuel bought ten acres and he and Alva grew a large garden. Alva did most of the work, planting, hoeing weeds, harvesting, and selling the produce. He didn't care for the job, saying that "hoeing corn in a hot sun is unattractive." Still his work helped support his family.

At age twelve, Alva decided it was time to start working at a regular job. After convincing his mother that he was old enough, he got a job working on the Canadian Grand Trunk Railroad. He didn't work for the railroad directly, but sold magazines, newspapers, and candy to customers on the train. As the train clacked down the track, Alva would walk the aisle selling the items in his basket. The train left Port Huron early in the morning, arrived at Detroit, sixty-three miles away, and then returned to Port Huron in the late evening. The schedule left Alva with several free hours in Detroit, but it was usually eleven o'clock at night before he arrived back home.

About the same time that Alva began his railroad job, he began to go deaf. Probably some of the cause was the scarlet fever he had as a child. Another cause he considered was an incident when he was late getting to the track and the train was already moving. As he ran to catch it, the conductor grabbed him by what he could reach—Alva's ears—and pulled him on board. Alva later recalled that "I felt something in my ears crack and right after that I began to get deaf." Years later, he wrote in his diary, "I have not heard a bird sing since I was twelve years old."

Detroit was a thriving, bustling town of 50,000 people. Alva could watch trains and ships, and visit machine shops. The shops provided chemicals for his laboratory. With the permission of Alexander Stevenson, the train conductor, Alva set up a laboratory in a baggage car. When he was not selling magazines and candy, he worked on his experiments. One day, an accident happened. Some phosphorus sticks were exposed to air and burst into flames. The fire seared the baggage-car floor and Stevenson burned his fingers putting out the flames. He forced Alva to remove his laboratory from the train.

In its place, Alva put a printing press and began his own weekly newspaper. The *Grand Trunk Herald* covered events happening up and down the track from Port Huron to Detroit. Alva sold them for three cents a copy. Setting type, especially on a moving train, was difficult and Alva ceased printing the paper after a few issues.

With the help of a friend, Will Wright, who worked at a newspaper, Alva continued in the newspaper business for a while. Alva handled the writing and editing, and Wright printed each issue of the paper called *Paul Pry*. The stories and gossip items included one story that so infuriated a Port Huron doctor that he threatened to throw Alva into the Saint Clair River.

It was 1862 when Alva began his first paper. Events in the world would bring him an opportunity that became his biggest success as a young entrepreneur. The Civil War was raging across the country. On April 6, he arrived at the offices of the *Detroit Free Press* to pick up the newspapers he would sell on the route back to Port Huron. He noticed a large crowd outside of the newspaper office. The people were waiting for word about a battle being fought in Tennessee. The battle, now called the battle of Shiloh, involved thousands of men, and their families anxiously awaited casualty

As an early teen, Alva, always industrious, bought a printing press
and started his own newspaper.

lists published by the paper and news of the battle. Alva saw a chance to make a lot of money.

Alva persuaded the editor of the *Free Press* to allow him to have a thousand papers on credit. He usually bought three hundred for the return trip. For several free issues of magazines and papers, he convinced a telegraph operator to telegraph the stations down the line. The announcement about the battle was written on blackboards at each station.

As the train entered each station, Alva sold many more newspapers than usual. At first, he sold them for five cents each. After seeing the demand, he raised his price to twenty-five cents. By the time he reached the end of the line at Port Huron, he had sold all the papers and made enough money to pay for them and a nice profit. The incident convinced him of the importance of selling what people wanted and of the value of the telegraph.

It was the telegraph that would lead him into the world of science and inventions in the years to come.

two
First-Class Telegrapher

Edison was fascinated with the telegraph. He and a friend built a telegraph line between their homes and practiced sending messages. His exposure to the telegraph operators along the railroad increased his curiosity about the device.

The telegraph was powered by electricity supplied by batteries. When the telegraph key was pushed, an electrical circuit was closed; releasing the key opened the circuit. In this way, tiny bursts of electricity flowed through the wires. At the receiving end, the electrical signals would enter an electromagnet and create a tiny magnetic field, activating a click from a sounding device.

Samuel Morse didn't invent the telegraph, but his devising of a code (named after him) allowed the telegraph to be used as an efficient means of communication. In Morse Code, dots and dashes stand for each letter and number. The letter

Samuel Morse, developer of the Morse Code *(Library of Congress)*

A is a dot, then a dash; B is a dash followed by three dots. The dots and dashes come from the space of time between clicks of the telegraph key—a short time is a dot, a longer time is a dash.

Fascinated by the communication device, Edison learned Morse Code, and often observed the telegrapher James MacKenzie at the Mount Clemens train station. One day, MacKenzie's three-year-old son was playing on the tracks as a freight train approached. Edison lunged onto the track, grabbed the boy, and rolled away from the train, saving the boy. In his gratitude, MacKenzie agreed to tutor Edison in telegraphy.

When Edison began his lessons, telegraphy was a thriving business. Telegraph wires were installed along the railroads across the country, allowing the transmission of news from coast to coast. The railroads employed telegraphers to handle messages of schedule changes, maintenance or track problems, and weather information.

The telegraph was making advances in other industries as well. Newspapers began to use the telegraph to send reports from far places quickly. In the financial districts of cities, monetary news was spread by telegraph. During the Civil War, the telegraph had proven itself as a military tool.

Edison practiced for hours and in 1863, he was hired as a part-time telegrapher in Port Huron. The telegraph office occupied a corner of Thomas Walker's jewelry store. Part of Edison's job was adding acid to the batteries that powered the telegraph. He also performed experiments in the back room of the store. One day, either an experiment went wrong or he added too much acid to the batteries, but a small explosion upset his employer enough that Edison was out of a job.

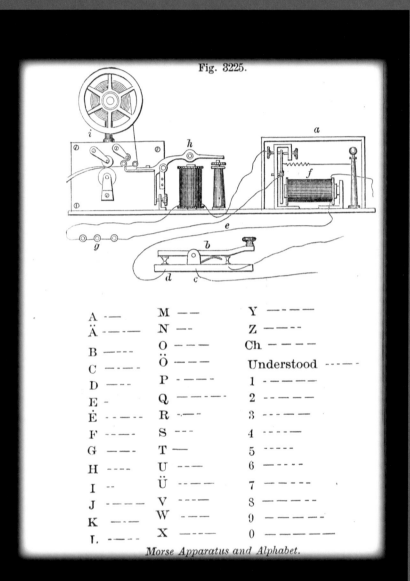

Fig. 3225.

Morse Apparatus and Alphabet.

A diagram of a telegraph machine along with a listing of Morse Code translations. *(Library of Congress)*

The Civil War years meant a high demand for telegraphers, so Edison quickly found another job. Like many others, he became a "tramp telegrapher," traveling from town to town, usually staying in one place only a few months. About this time, he began going by his first name, dropping the Alva or Al that he had been called through his childhood.

As Edison traveled to jobs in Michigan, Ohio, Tennessee, Indiana, Kentucky, and Canada, he packed gear to continue his experiments. Any money left from living expenses went to buy more electrical equipment, wire, or chemicals.

Most telegraphers of the time were hard living men, living in run-down boardinghouses and reveling in vices such as drinking and chewing tobacco. The telegraph offices were usually ramshackle buildings with little comfort for the employees. Edison wasn't much of a drinker; one drink would put him to sleep. But he excelled as a practical joker. One time, he wired an induction coil to a water bucket. Each man who dipped his cup in to get a drink received an electric shock. In Cincinnati, he wired a hand-washing sink to do the same thing. His fellow employees did not always appreciate his jokes.

Sometimes his actions got him fired. During an experiment in the Louisville office, he dropped a jar of battery acid. The acid ate through the floor and ruined furniture in the office below.

Another incident was potentially more dangerous. While working at Stratford Junction, Canada, Edison received a telegraph to hold a freight train and he ran out to find the signalman. Before he could find the signalman, the train rushed by. Edison ran back to telegraph that he had been unable to hold the train. The dispatcher had let another train enter that track from the opposite direction. Edison began stumbling through the dark to reach the next telegraph station and stop the freight train. He fell into a culvert and didn't make it to the station in time. Fortunately, the track was straight and the two engineers saw each other. They were able to stop the trains before crashing into each other. Edison and his supervisor were called to the main office to answer for their actions. Edison managed to slip away and sneak into a train leaving Canada into the United States, just as his father had.

He did make some friends among the telegraphers. He, Milt Adams, and Ezra Gilliland attended operas and plays together and sometimes shared a room. Despite being partially deaf, Edison shared his friends' appreciation of music; he especially liked the music of Beethoven. He said, "Beethoven's greatest symphonies are the summit of human creativity."

Being hard of hearing was a blessing in his telegraph jobs, according to Edison. He figured that his deafness prevented him from hearing much of the noise around him, so he could concentrate fully on what he needed to hear from the telegraph receiver.

Edison's curiosity led him to experiment with the telegraph machines. In Indianapolis, he found an old Morse code machine that was used before operators received messages by ear. The machine recorded the code on a strip of paper. Edison devised a way to regulate the speed of a second machine, so when he ran the coded strip through it, he could take as long as needed to transcribe the code.

He tried to use this method to send messages without relaying between operators. For instance, a message could only travel two hundred miles with the power provided by batteries. Then an operator would have to copy the message down and retransmit it. This system allowed errors to creep into the messages. Edison thought that by using his repeating machine, the message could be retransmitted error free. Unfortunately, he was fired for his idea. His supervisor had been working on a similar idea and didn't want Edison competing with him.

Another idea that he was working on was the duplex telegraph. Messages could only be sent one at a time in one direction. Edison saw that if he could make the wire carry a message in each direction he would double the capacity of the wire. His boss didn't agree. He said, "Any damn fool ought to know that a wire can't be worked both ways at the same time." In a few years, Edison would be sending not two, but four messages on the same wire.

Edison wanted to be classified as a "first-class man," meaning a top-speed telegrapher. Such a person could send or receive messages at twenty-five to forty words a minute. He found a way to increase his receiving speed—writing in tiny letters.

In the fall of 1867, Edison went for a visit with his parents in Port Huron. Samuel and Nancy were still having financial problems and had moved into a smaller house. Nancy also battled depression. Edison became ill that winter and ended up staying several months.

When he began looking for a job again, he remembered his friend Milt Adams, who had moved to Boston. Edison wrote to see if Adams knew of jobs and the reply said yes, come as soon as you can. Edison repaired a telegraph line for his old employer, the Grand Trunk Railroad and received a ticket as pay.

It was January 1868, and the train route took it up into Canada. A blizzard stranded Edison on the train and it took four days to reach Montreal. He finally arrived in Boston, ready for his job with the biggest telegraph company, Western Union.

The Boston telegraphers made fun of his country ways, his speech, his flannel shirt, and too short jeans. Then they arranged an initiation test for him. Edison was ordered to take the press line, meaning he would receive the Morse code of news copy from New York. The Boston and New York offices competed to have the fastest senders and receivers. As a man named Hutchison from New York began sending

This photo of the New York telegraph office shows how extensive the telegraph business was by 1875. *(Library of Congress)*

code, Edison realized that he was extremely fast. With his tiny writing, Edison was able to keep up, even stopping at one point to sharpen his pencil. Finally, Edison ended the contest by sending Hutchison a message: "You seem to be tired, suppose you send a little while with your other foot." Winning the contest made Edison a hit at the Boston office.

Edison soon got into trouble with his supervisors. One of his tricks for receiving code so quickly was to write letters tiny enough to read with a magnifying glass. When his supervisor complained, Edison wrote a message in letters so large that he used several pieces of paper. He was demoted for this prank.

He also repeated his stunt with the water pail and another telegrapher became so angry that he threw a heavy glass insulator at Edison's head. Edison managed to duck. Another prank was gluing the telegraph keys together.

Edison enjoyed the advantages of Boston. The city was the most technological place in the country at the time. Inventors and tinkerers flourished in the city as the Industrial Revolution roared along. Edison and other telegraphers gathered to talk shop and dream of inventions at

Edison in 1870, around the time he received his first patent for the voice recorder. *(Courtesy of Hulton Archive/Getty Images)*

Charles Williams's store; Williams sold telegraphic equipment and was an inventor.

At a used bookstore, Edison bought a three-volume set of Michael Faraday's books on electricity: *Experimental Researches in Electricity*. After reading Faraday's clear

Michael Faraday

experiments and conclusions about electricity, magnetism, and their connection, Edison had a new hero along with Thomas Paine. Referring to Faraday, he told his roommate, Milt Adams, "I am now twenty-one. I may live to be fifty. Can I get as much done as he did? I have got so much to do, and life is so short. I am going to hustle."

An Englishman, Faraday's tinkering with electricity and magnets had led him to the invention of the dynamo and the principle of electromagnetism. He had died the year before Edison arrived in Boston, but his discoveries were transforming the world. Inspired by Faraday's work and creativity, Edison set out to become an inventor in his own right.

Edison's first significant invention was a vote recorder. In Congress or state assemblies, votes were tallied by voice. Each official was required to say yes or no, and the effort of polling everyone often took hours. Edison's vote recorder could accomplish the task in a few minutes. At each member's desk would be two buttons; one for yes and one for no. The voter would press their chosen button, and the signal traveled through wires and pushed a marker on the tally board.

Edison applied for a patent on the vote recorder in November 1868 and it was awarded in June 1869. It was his first patent and over his lifetime, he would receive 1,093, more than any other person. He demonstrated the vote recorder to the legislature of Massachusetts and to Congress in Washington, DC.

Unfortunately, he couldn't sell any vote recorders. Members of state assemblies and Congress did not want a faster way to record votes. They used the time during a voice roll call to campaign for votes and make arguments for or against a proposal. The telegraphic vote recorder was a failure, and

Edison resolved to never again invent something that the public didn't want.

Edison continued his experiments in hours after work at the telegraph office. He and others were still working on a duplex telegraph, one that would send two messages at a time. An article about Edison's work was published in an industry newsletter, the *Telegrapher*. Edison became friends with the owners of the journal, Franklin L. Pope and James Ashley.

He also became interested in a device for the financial industry, a stock ticker. The machine would report gold and

An Edison stock ticker *(Courtesy of AP Images)*

stock prices instantaneously to a receiver that made a ticking noise as it printed the prices on a strip of paper.

His friend Pope worked for a gold-reporting company and Pope had made improvements in the stock ticker. Edison improved it even more by using less wire in the machine. He and others set up a gold reporting service in Boston and signed up several companies. The machines were installed in the offices of the subscribers and they received regular updates on the price of gold.

In January 1869, Edison quit his telegrapher job with Western Union. In a notice in the *Telegrapher,* he announced, "T. A. Edison has resigned his position in the Western Union office, Boston, and will devote his full time to bringing out inventions."

three
Beginning Inventor

One of Thomas Edison's first acts as an independent inventor was to test his duplex telegraph. Franklin Pope got permission to conduct the test on lines of the Atlantic and Pacific Telegraph Company. Edison and Pope tried to send messages to each other on the line stretching from New York City and Rochester, New York. The test was not a success. Edison had trouble getting the equipment he needed, the lines had poor insulation, and Edison ran out of time.

His debts were growing and with the failure of the duplex test, he needed to find some way to raise money. The gold stock company was not doing well in Boston, so Edison borrowed money for a steamship ticket to New York. He hoped the financial community in New York City would be less tight on money than the conservative Boston had been.

When he arrived in New York, he had almost no money and was hungry. He talked a teashop owner into letting him have a sample package of tea, which he traded to another shop for a breakfast of baked apple dumpling and coffee. He intended to stay with a friend, but he wasn't home, so Edison slept on the streets the first night.

The next day, he contacted Pope. Besides, part owner of the *Telegraph* newsletter, Pope worked for Samuel Laws at a gold-price information service. The company served subscribers in the New York area, sending them gold prices as they changed—just like Edison's gold price company in Boston. No jobs were available at the company, but Edison was allowed to hang around and sleep in a workroom.

One day when Pope was out of the office, disaster struck. The telegraphic transmitter stopped. Subscribers from all over the financial district were sending messengers to find out why they were not receiving the gold price updates. Chaos reigned in the office because no one knew how to fix the equipment. Edison inspected it and found that a spring had broken and jammed the transmitter. "Fix it! Fix it!" Laws yelled at him and in about two hours, Edison had replaced the spring and the transmitter started working.

Then there was another problem: in each of the subscriber's offices, the receivers had to be reset by hand, taking a long time. Edison told Laws that he could invent a way to stop all the receivers instantaneously if the transmitter stopped. Then they could all be started together again without having to be manually reset. Laws was impressed enough to offer Edison a job as Pope's assistant.

The job paid $225 a month. Soon after Edison began, Pope left the company and Edison became the manager. But it didn't last long. In a few months, Laws sold the company to his competitor, the Gold & Stock Telegraph Company. They had their own manager, so Edison was out of a job. One of the reasons the Gold & Stock Telegraph Company wanted Laws' company was to control Edison's patent on the automatic stop mechanism. His invention had cost him a job.

Always optimistic, he was not discouraged but saw an opportunity. He and Pope formed their own company. An announcement on October 2, 1869, in the *Telegraph* said "Pope, Edison & Co., Electrical Engineers and General Telegraphic Agency." They offered design and manufacture of telegraphic equipment, including new inventions; construction and maintenance of telegraphic systems; testing of systems; and supplies. "A leading feature will be the application of Electricity to the Arts and Sciences," the ad read. James Ashley, the other owner of the *Telegraph,* was a silent partner in the company.

They rented spaces in Jersey City, New Jersey and set up their workshop. Edison lived with Pope's parents nearby.

The main customer for Edison and Pope was not the big telegraphic companies, such as Western Union. Those institutions strung their lines and transmitted messages across long distances, from town to town and state to state. But in large cities, including New York and Boston, urban telegraph companies were stringing wire everywhere. These telegraphs were private lines connecting offices to customers or warehouses and offices to financial concerns.

Soon, the trio of Edison, Pope, and Ashley began another company—Financial and Commercial Telegraphy Company. The equipment promoted by the company was a printing telegraph designed by Edison and Pope. The new machine put them in competition with Gold & Stock, Edison's former employers.

In the spring of 1870, Gold & Stock was feeling the competition from Edison, Pope, and Ashley's Financial and Commercial Telegraphy Company. The president of Gold & Stock, Marshall Lefferts, made a merger offer, mostly to get patent rights to Edison's inventions. For a payment of $15,000, Gold & Stock bought Financial and Commercial Telegraphy Company. However, Edison, Pope, and Ashley retained rights to private-line telegraphy.

They formed a new company to provide a printing telegraph.

Edison as a young inventor
(Library of Congress)

Lefferts was also part of the new company, American Printing Telegraph. The printing telegraph allowed inexperienced operators to send and receive messages without using Morse code, a needed improvement for business communications. The user-friendly machine won the "Best Electric Printing Telegraph Instrument" at the American Institute fair in New York.

Soon, Lefferts and Gold & Stock Telegraph also bought out American Printing Telegraph. During the negotiations, Edison, Pope, and Ashley had a falling out. Ashley had once praised Edison as "a young man of the highest order of mechanical talent, combined with good scientific electrical knowledge and experience." He now banned any mention of Edison in the *Telegraph*. Years later, he wrote of Edison as a "professor of duplicity and quadruplicity," implying that Edison stole ideas from other inventors and promoted them as his own.

This pattern of friendship and then animosity was to follow Edison through many relationships. He also became embroiled in numerous lawsuits, usually because Edison was willing to accept money for his inventions, even if he received support from two competing companies. Other inventors claimed that he stole their inventions, but many times he improved on their originals.

A relative of Lefferts, William Unger, joined Edison to make stock tickers for Gold & Stock. They opened a machine shop in Newark, New Jersey.

Edison also entered a partnership with George Harrington, an investor with the Automatic Telegraph Company. They set up a new company, American Telegraph Company, and Edison opened another shop in Newark for this work.

The investors wanted Edison to improve an automatic telegraph design invented by Englishman George Little. The company had patent rights to the device, but there were several flaws in the design.

On an automatic telegraph, the operator typed in code with a keyboard, and a stylus punched the code on a piece of chemically treated paper. The paper fed into a transmitter that could send messages at speeds of one hundred words per minute, more than twice as fast as a telegrapher.

In May 1870, Edison wrote to his parents. He had learned, probably from a cousin, that his mother was not well. In his letter, he told his father, "Dont do any hard work and get mother anything She desires=You can draw on me for money."

On October 30, he wrote another letter to his parents. "I may be home some time this winter=Can't say when exactly for I have a Large amount of business to attend to. I have one shop which Employs 18 men, and am Fitting up another which will Employ over 150 men=I am now what 'you' Democrats call a 'Bloated Eastern Manufacturer.'"

Edison was twenty-three years old when he wrote this letter. Most of the men that he employed were older than he was, but he called them his "boys." They called him the "Old Man." Edison and his employees worked long hours, sometimes napping on the floor or a table.

Tensions soon developed between Edison and Harrington. The work on the automatic telegraph was proceeding too slowly for Harrington, and Edison was quickly spending the money invested in the company. After Harrington visited the shop, Edison wrote a friend and fellow investor, Daniel Craig. "Mr H says that some of our experiments are useless. But after he has had more experience in this business, he will find that No experiments are useless."

Still dissatisfied with Edison's management of the company, Harrington hired a shop supervisor. Edison immediately resigned.

In the midst of this turmoil, Edison received a telegram from Port Huron. His mother, Nancy, had died on April 9. She had struggled with an unknown disease that affected her mind. Edison traveled home for her funeral and then returned to Newark and his work.

Later in 1871, Edison met a young lady named Mary Stilwell. She worked as a transmitter operator for Gold &

Stock Telegraph. Edison still conducted quite a lot of business with the company, so he had many occasions to meet Mary. According to one account, one day Mary said, "Mr. Edison, I can always tell when you are behind or near me."

"How do you account for that?" he asked.

"I don't know, I am sure," she answered, "but I seem to feel when you are near me."

Edison replied, "Miss Stilwell, I've been thinking considerably of you of late, and if you are willing to have me I'd like to marry you. . . . I know you never thought I would be your wooer, but think over my proposal, Miss Stilwell, and talk it over with your mother. Let me know as soon as possible, as if you consent to marry me, and your mother is willing we can be married by next Tuesday." Mary's mother consented and the lightning fast courtship was over.

Thomas Edison and Mary Stillwell were married on Christmas Day, 1871. He was twenty-four, she sixteen. Mary came from a working-class family and began working early herself, but her marriage to Thomas raised her social standing. She had little interest, however, in his work. In a notebook entry dated February 14, 1872, Edison wrote, "My Wife Popsy Wopsy Can't Invent."

Edison and Mary had rough times in their marriage. His absorption with work continued, so he was often gone sixteen hours a day or even days at a time. Mary would be lonely and neglected, so she went shopping. She was no better at handling money than her husband, so their financial status was never stable.

In February 1873, they had a daughter. They named her Marion, but Edison quickly gave her a nickname—Dot.

Even with his new family, Edison's lifestyle changed little.

He smoked a pipe and occasionally chewed tobacco. When he was in the midst of work, a meal would often be apple pie and coffee.

Edison's offhand dealing with others and with contracts kept him in conflict with his partners and investors. In 1872, he and Unger split ways, and Edison went into debt to buy Unger's part of the Newark shop. He renewed his relationship with Automatic Telegraph Company and sailed to England to show the automatic telegraph system to the British Post Office. The British were impressed by the tests of the system, but decided not to buy it. Still, the years of 1872 and 1873 were fruitful for Edison. He filed for nearly sixty patents. Many of the patents were not brand new inventions, but improvements on existing products. Edison was becoming well known on telegraphy.

He also reconnected with Western Union by promising the president, William Orton, that he could build a quad telegraph, meaning four messages sent at a time. Orton agreed to let him use the Western Union shop to work on the invention.

In 1873, a panic hit the country, brought on by the failure of banks and a plunge in stock values. Edison owed money and other people owed him, but no one could pay. Edison was forced to sell the house where the family lived and move into a small apartment.

He asked Orton for more money for his experiments on the quad telegraph. As they were negotiating, Orton left town on a business trip. While he was gone, Edison met with a competitor of Western Union. Jay Gould, president of Atlantic & Pacific Telegraph Company, wanted to buy Automatic Telegraph Company to try to corner more of the telegraph market away from Western Union. He showed interest in

Edison's quad idea and offered Edison $30,000 in cash and company stock for the rights to any patents that Edison might obtain for the quad system. Edison had a verbal agreement with Orton, but he decided to take Gould's deal. He would also work for Gould's company.

A complicated legal battle erupted from these dealings. Western Union claimed rights to the quadruplex and so

Jay Gould *(Library of Congress)*

Gould backed out of the stock deal with Edison and his fellow partners in Automatic Telegraph. Edison and the others sued Gould. The legal wrangling went on for years. In 1881, Jay Gould seized control of Western Union.

With the money he got from Gould, Edison moved Mary and Dot into a new house. He brought his father and a nephew, Charley, to live with them. He also started a laboratory to provide space to work on his inventions and study other sciences, such as chemistry.

Edison had two new employees in the invention business. One, Charles Batchelor, was born in England. He was a skilled machinist and draftsman and became Edison's most trusted lieutenant. Edison also hired John Kruesi, a Swiss-born machinist who could make any machine that the inventors designed.

One intriguing invention coming from this lab was the electric pen. A minute electric motor powered by a battery caused a tip to vibrate, perforating waxed paper as a person wrote. The paper could then be used as a stencil to make copies of what had been written.

In November 1875, Edison made a remarkable discovery. With the urging of William Orton of Western Union, he was investigating the idea of an acoustic telegraph. This method would split telegraphic signals into tones and send them over the wires. More messages could possibly be sent this way than even Edison's quadraplex system.

During sound experiments using a vibrating magnet, Edison and some of his employees noticed brilliant sparks shooting out of the core of the magnet. They experimented with these sparks and found that they could be sent around the metal pipes of the laboratory. The strange sparks did not register

Charles Batchelor *(Library of Congress)*

Edison's electric pen (*Courtesy of AP Images*)

on a galvanometer, an instrument to measure electrical current, but they could pass through insulating materials such as rubber and glass. Edison concluded that what they saw was a new, unknown force and he named it etheric force.

Edison wrote some articles about his discovery, but scientists laughed at him. One physicist, Dr. George M. Beard, believed him and hypothesized that Edison had found "a radiant force, somewhere between light and heat on the one hand and magnetism and electricity on the other." Edison saw no practical applications for this force, so he returned to his inventions.

What Edison had observed was high-frequency electromagnetic waves. These waves include light, x-rays, ultraviolet

Heinrich Rudolf Hertz used Edison's inadvertant discovery of high-frequency electromagnetic waves to begin the development of the radio.

waves, and radio waves. In the middle 1880s, German scientist Heinrich Rudolf Hertz experimented on these waves and his work became the foundation for the development of radio.

Up until this time, most of Edison's inventions had been improvements or new equipment for the telegraph. But he was soon to move into a new place of business and a new arena of inventions.

four
Sound
Invention

By the end of 1875, Edison was planning to relocate his laboratory and workshop. He was looking for a quieter place with cheaper land prices, so he would have plenty of expansion room and space for a home. His daughter, Dot, would soon be joined by another child.

One of his employees told him about a tiny community located twelve miles south of Newark, New Jersey. Menlo Park only had six houses, but a railroad line passed nearby giving quick access to New York City. Edison was able to buy two pieces of land and a house for $5,200.

He put his father in charge of construction of the new laboratory at a cost of $2,500. The laboratory building was one hundred feet long and thirty feet wide. The first floor contained a library, drafting room, office, and machine shop. The second floor was a large laboratory, described by a writer for *Popular Science Monthly*: "The walls are covered

Edison's home at Menlo Park *(Library of Congress)*

with shelves full of bottles containing all sorts of chemicals. Scattered through the rooms are tables covered with electrical instruments . . . microscopes, spectroscopes, etc. In the centre of the room is a rack full of galvanic batteries."

Edison's goal for the new research lab was "a minor invention every ten days and a big thing every six months or so." The employees worked as a team with Edison as chief of what he called his invention factory. The combination of laboratory and workshop employing many people was the beginning of research laboratories. Most large companies today have their own research and development departments, similar to Edison's Menlo Park.

Edison moved into the new lab and home in March 1876. His second child, a son named Thomas Alva Jr. was born in

January 1876. Edison nicknamed him Dash. With his daughter nicknamed Dot, Edison had the whole telegraphic alphabet. But with his attention mostly on the lab, Edison's wife and children rarely saw him. He ate most meals at the lab and even slept there often, stretched out on a desk or the floor.

An article in the *New York Herald* described the routine:

> Edison and his numerous assistants turn night into day and day into night. At six o'clock in the evening the machinists and electricians assemble in the laboratory. Edison is already present, attired in a suit of blue flannel, with hair uncombed and straggling over his eyes, a silk handkerchief around his neck, his hands and face somewhat begrimed and his whole air that of a man with a purpose and indifferent to everything save that purpose. By a quarter past six the quiet laboratory has become transformed into a hive of industry. . . . Edison himself

A magazine illustration of the Menlo Park laboratory in its early days. *(Library of Congress)*

flits about, first to one bench, then to another, examining here, instructing there, at another earnestly watching the progress of some experiment. Sometimes he hastily leaves the busy throng of workmen and for an hour or more is seen by no one. Where he is the general body of assistants do not know or ask, but his few principal men are aware that in a quiet corner upstairs in the old workshop, with a single light to dispel the darkness around, sits the inventor, with pencil and paper, drawing, figuring, pondering. In these moments he is rarely disturbed. If any important question of construction arises on which his advice is necessary the workmen wait. Sometimes they wait for hours in idleness, but at the library such idleness is considered far more profitable than any interference with the inventor while he is in the throes of invention.

The move to Menlo Park provided a quiet atmosphere for Edison's inventions, but it was too quiet for Mary. She was more isolated than ever, and reached out by splurging on lavish furnishings, causing even more financial instability for the family.

Edison was still working on the acoustic telegraph. Two other inventors were also experimenting in the field: Elisha Gray from Chicago had sent eight messages at a time using his form of the acoustic telegraph; and Alexander Graham Bell from Boston was performing sound experiments to send not musical tones, but the voice. The instrument that they were inventing became the telephone. The United States Patent Office concluded that Bell's invention came first.

On March 10, Bell and his assistant, Thomas Watson, conducted their famous experiment. With wires between their transmitter and receiver running from room to room, Bell spoke into the transmitter, "Mr. Watson, come here; I

An 1882 portrait of Alexander Graham Bell *(Library of Congress)*

want to see you." Watson heard the words, even though the sound was very faint.

By the fall, Bell could project a voice over several miles, an exciting development. Still there were problems. The main one was that a person had to yell into the transmitter to be heard.

Edison was conducting his own sound experiments. He zeroed in on perfecting the transmitter so that a person would not have to shout into it. A telephone contains a diaphragm, or a flexible disk, that vibrates in response to sound waves such as speech. The vibrations are converted into electrical signals which travel down the wire and then are changed back into sound by the receiver.

Edison and Batchelor performed several experiments to improve the sound quality. As he often did, Edison took an idea from one invention and used it for a different one. In his automatic telegraph, he had used a carbon disk that regulated the amount of current passing through it. Through the fall and winter of 1877, Edison and his employees performed more than 2,000 experiments with carbon compounds. The best solution came from a carbon button placed next to the diaphragm. Putting the carbon disk in the transmitter of the telephone improved the sound of speech. Edison sold his carbon-button transmitter to Western Union. Many telephones today still have carbon buttons.

To speak on Bell's telephone, a person spoke into the transmitter and then turned it around to hear from the receiver. Edison combined the two functions into one piece, similar to today's handset. He applied for a patent on his telephone work in April 1877, but fifteen years of lawsuits with Bell passed before Edison got a patent for his telephone work.

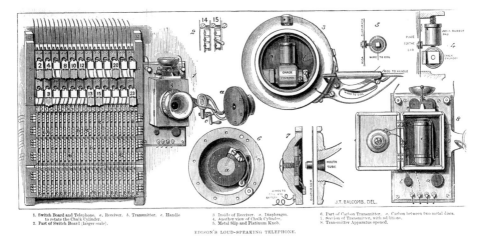

1. Switch Board and Telephone. *a.* Receiver. *b.* Transmitter. *c.* Handle to rotate the Chalk Cylinder.
2. Part of Switch Board (larger scale).

3 Inside of Receiver. *a.* Diaphragm.
4. Another view of Chalk Cylinder.
5. Metal Slip and Platinum Knob.

6. Part of Carbon Transmitter. *a.* Carbon between two metal discs.
7. Section of Transmitter, with additions.
8. Transmitter Apparatus opened.

EDISON'S LOUD-SPEAKING TELEPHONE.

A diagram of the different components of Edison's improved telephone *(Courtesy of Mansell/Time Life Pictures/Getty Images)*

Bell expected people to answer the telephone with the word ahoy. But Edison's team came up with the word hello. It originated from a word for attracting attention—halloo— and they changed it to the word used today.

As Edison considered the experiments with the telephone, he began to think about recording speech. He thought that his main client, Western Union, could record a message and then deliver the medium on which the message was recorded. His ideas were leading to a talking machine.

He attached a pin to a diaphragm and put a piece of waxed paper under it. "I rigged up an instrument hastily and pulled a strip of paper through it, at the same time shouting, 'Halloo!'" When he pulled the marked paper under another diaphragm and pin riding the groove created, he and Batchelor listened carefully. They heard a sound that Edison said might be imagined to be "Halloo!"

Edison tried several types of materials to receive the impression made by the pin and diaphragm, including wax, chalk, and tinfoil. At the end of November 1877, he gave some sketches to master machinist John Kruesi, and asked him to build the instrument. When Kruesi asked him what he was building, Edison answered, "The machine must talk."

Kruesi built the machine; a rotating cylinder turned by a hand crank and a diaphragm and needle at each end. On December 6, Edison wrapped a piece of tinfoil around the cylinder and leaning close to the machine, recited "Mary Had a Little Lamb" into it as he turned the crank. The needle inscribed a groove on the cylinder. He placed the needle of the second diaphragm into the groove left by the first needle, turned the crank, and to everyone's amazement, Thomas Edison's voice was heard quoting the nursery rhyme.

The rest of that night, Edison and his team played with the machine that he decided to call a phonograph. The men sang and talked into it and then listened to their recordings. Each cylinder would only hold a couple of minutes of sound. During the night, they made a second model as Batchelor recorded in his notebook, "Finished the phonograph. Made model for P.O. [Patent Office]."

The next day, December 7, Edison and Batchelor were in Washington, DC. They demonstrated the phonograph to Alfred Beach, editor of *Scientific American*. Edison put the machine on the editor's desk and played a recording he had made. The editor wrote: "the machine inquired as to our health, asked how we liked the phonograph, informed us that *it* was very well, and bid us a cordial good night. These remarks were not only perfectly audible to ourselves, but to a dozen or more person gathered around, and they were produced by the aid of no other mechanism than the simple little contrivance." The other employees of the magazine congregated to hear

Edison's original tin foil phonograph *(Library of Congress)*

the machine, and so many people were there that the editor feared that the floor of his office would collapse.

The phonograph became a sensation. Many were surprised that the phonograph had been invented by someone who was partially deaf. Articles appeared in newspapers and magazines, and soon people were showing up in Menlo Park to see the famous Edison and his lab. So many visitors were coming for a glimpse of the celebrity inventor that it was hard for him to get any work done. Those who did see him observed a man with untidy hair, dressed in rumpled clothes—usually a worn black coat spotted by chemicals. He would talk to them in his flat midwestern speech, answering questions. One man said, "I understand it all, except how the sound gets out again." Eighty to one hundred letters were arriving daily.

Edison was called the "Inventor of the Age" and the "Wizard of Menlo Park." A cartoon describing him as the

Wizard shows him wearing a tall hat marked with scientific symbols and a robe covered with technical drawings.

On April 18, 1878, Edison demonstrated the phonograph to the United States Congress and to the National Academy of Sciences. He was also invited to the White House by President Rutherford B. Hayes, where he thrilled the president. Edison recalled, "The exhibition continued till about 12:30 A.M., when Mrs. Hayes and several other ladies who had been induced to get up and dress, appeared. I left at 3:30 A.M."

Edison and his team kept improving the phonograph, substituting a wax cylinder for the tinfoil one and experimenting with spring loaded or motorized cranks. He started a company, Edison Speaking Phonograph Company, to manufacture and market phonographs. At first, they were so tricky to operate that trained men operated them in exhibitions;

The Phonautograph

In March 2008, American researchers discovered that Thomas Edison was not the first to record sound. That distinction goes to a little-known French inventor named Edouard-Leon Scott de Martinville, who lived in Paris and worked as a typesetter and librarian.

On April 9, 1860, Scott made a ten-second recording of a vocalist singing a French folk song on a phonautograph—a device created by Scott that etched sound waves onto sheets of paper blackened with soot from an oil lamp. His recording predates Edison's recording of himself reciting "Mary had a little lamb" by nearly two decades.

The discovery was made by a team of U.S. audio historians, recording engineers, and sound archivists on a quest to find and compile an anthology of the world's oldest recorded sounds. They unearthed the recording, described as "a historic find, the earliest known recording of sound," in Paris, along with recordings Scott made in 1853 and 1854. However, the 1853 and 1854 recordings, or phonautograms, turned out to be little more than a "squawk."

Scott never envisioned his phonautograph as a device to play or reproduce any sounds. Instead, he created it to make a visual image of sound waves. The phonautograph had a barrel-shaped horn attached to a hog's bristle stylus that etched sound waves onto sheets of smoke-blackened paper.

Scientists at the Lawrence Berkeley National Laboratory in Berkeley, California, made high-resolution digital scans of the soot-covered paper on which the French folk song was recorded and then used a "virtual stylus" to read and convert the lines into sound.

The recording is of a person, perhaps a young woman, singing "*Au clair de la lune, Pierrot repondit* ("By the light of the moon, Pierrot replied").

Scott was aware of Edison's invention of the phonograph, and he reportedly scorned him for "appropriating" his techniques. In a self-published memoir in 1878, Scott wrote, "What are the rights of the discoverer versus the improver? Come, Parisians, don't let them take our prize."

There is no evidence, however, that Edison knew about Scott's work. He made his advances independent of Scott, and still today, Edison retains the honor of being the first person to have recorded sound *and* played it back.

Edison and his phonograph during his visit to Washington, DC, in 1878. *(Library of Congress)*

people would pay twenty-five cents to hear a recording of a few minutes. The phonograph was added to exhibits that Edison was sending to the Paris Universal Exposition.

Edison hoped the phonograph would be used for many things besides entertainment. He thought it would be useful for recording family members, teaching speech and diction, preserving the voices of famous people, speaking for the blind, and in the business world, as a dictation device. He always claimed that the phonograph was his favorite invention. He told a journalist "this is my baby and I expect it to grow up to be a big feller and support me in my old age." He filed for a patent on December 15, 1877, and received it two months later.

By summer of 1878, Edison was exhausted. Professor George Barker of the University of Pennsylvania invited him to go west for a special event—a total eclipse of the sun. Edison had just invented what he called a tasimeter, an instrument to measure minute changes in temperature, as small as a millionth of a degree Fahrenheit. The eclipse seemed like a good time to test it. Before Edison departed for Rawlins, Wyoming, from the Pennsylvania Railroad station, he told a reporter "I can hardly wait until I get there. This is the first vacation I have had in a long time, and I mean to enjoy it. I have never seen the country. . . . After we have made the observations, however, I am going further West, to take in the Yosemite, San Francisco and a portion of the coast."

The eclipse occurred on July 29. Edison prepared the tasimeter, but the weather was against him. High winds caused vibrations that threw the instrument out of adjustment. Still, "the experiment had shown the existence of about fifteen times more heat in the corona [of the sun] than that obtained from the star Arcturus the previous night."

Edison's trip west lasted two months. The country he traveled through was still relatively unsettled compared to the East. Wyoming was still a territory and would not gain statehood for twelve more years. Only two years earlier, General George Armstrong Custer had been defeated by the Sioux Indians at the Little Big Horn in Montana. Edison and his companions traveled through Wyoming, Nevada, Utah, and into California. During the trip Edison enjoyed fishing, hunting, and camping.

Edison convinced the railroad company to let him ride on the cowcatcher to have an unobstructed view of the countryside. Trains then traveled at about twenty miles per hour, but Edison traveled at his own risk. Edison later told the story: "only once was I in danger when the locomotive struck an animal about the size of a small cub bear, which I think was a badger. This animal struck the front of the locomotive just under the headlight with great violence and was then thrown off by the rebound. I was sitting to one side grasping the angle brace so no harm was done."

San Francisco and Yosemite were stops on Edison's journeys. In Nevada, he visited large silver and gold mines and thought about how electricity might be used to mine these precious metals.

While Edison was traveling, Mary was dealing with a difficult pregnancy. He received a telegram that she was ill, but he decided to continue on to St. Louis where he planned to attend the American Association for the Advancement of Science (AAAS) Meeting and then he would be home. Apparently, he figured that his wife was being a hypochondriac. She had been ill a lot and was considered high-strung and nervous.

At the AAAS meeting, Edison was introduced by George Barker who commented, "the time has come when the scientist is no longer the only discoverer; the practical man has found science too slow, and has stepped in and discovered for himself."

Edison returned to Menlo Park on August 26, refreshed and ready to return to inventing. He had a field trip planned for September, arranged by George Barker. He and Batchelor traveled to Ansonia, Connecticut, to visit a foundry owned by William Wallace. The brass and copper foundry was lit by the newest in electricity—arc lights. Wallace had built several dynamos and hooked one to a turbine powered by a nearby waterfall. The electric current produced was much stronger than what could be obtained from batteries. The current powered a series of arc lights. The lights had carbon plates or rods, and an electric current jumped the gap between the rods. The arc light was a brilliant glow, but it was only suitable for large spaces or outdoor lighting. Poisonous gases were emitted as the carbon rods burned and the light was too bright for homes. Some cities in Europe were using arc lights for streets.

Edison was inspired by Wallace's foundry. The reporter that was with him wrote, "He ran from the instruments to the lights and from the lights back to the instrument. He sprawled over a table with the simplicity of a child, and made all kinds of calculations. He estimated the power of the instrument and of the lights, the probable loss of power in transmission, the amount of coal the instrument would save in a day, a week, a month, a year, and the result of such saving on manufacturing."

There had been arc lighting experiments at Edison's lab, but after his reviving trip west and the inspiring visit to the foundry, he was ready to tackle the problems of electric lights.

five
Light
Inventions

T homas Edison was not the first inventor or scientist to try to create the incandescent light bulb. In the early 1800s, England's Sir Humphry Davy had shown that electric current flowing through some materials causes them to glow or incandesce. For more than fifty years before Edison, scientists had tried to encase a filament in a glass ball and by applying current, make it glow. They found that most materials would glow for a short time and then either melt or burn up.

Edison had conducted some light experiments in 1877, but then became occupied with the invention of the phonograph. He later described his first experiments and why they did not work:

> [W]e were trying to subdivide the electric light into a small number of burners, where the circuit was closed by solid

conductors, and the reason why experiments were conducted with boron and silicon was because they were not subject to oxidation like carbon, which we had previously tried, and which did not last as long at a white incandescence as pieces of graphitoidal silicon. The results of the carbon experiments, and also of the boron and silicon experiments, were not considered sufficiently satisfactory, when looked at in the commercial sense, to continue them at that time, and they were laid aside.

After his visit to Wallace's foundry, Edison was inspired to try again to develop an incandescent light bulb. His plan called for not just electric lights in businesses and homes, but also electric machines run by a central power station. He knew that for an electric light system to be wanted by the public, it would have to be competitive in price with gas lights.

Gas lights were common in cities where the concentration of people made them practical. Coal would be turned into gas at a central plant and the gas pumped through underground pipes to businesses, factories, and homes. Each gas fixture had to be turned on and lit with a match and turned off and blown out. The burning of the gas threw off toxic vapors and there was always a danger of fire. The light produced by the gas was fairly bright, but flickered.

In a newspaper article dated September 16, 1878, Edison laid out his plan to create a lighting system for one square mile of lower Manhattan. His electricity would be produced by generators, flow through underground conduits, much like gas, and light homes and businesses in that area. He thought that gas light fixtures could be converted to electric lights.

Years would pass before his first lighting station was in operation in New York City, but his confident announcement led investors to talk to him about forming a company.

In November 1878, the Edison Electric Light Company was incorporated. Investors besides Edison included the president of Western Union, Norvin Green, financier Robert Cutting, Jr., Egisto Fabbri, a member of the Drexel-Morgan Company, J. P. Morgan, and others.

Attorney Grosvenor Lowrey acted as go between for Edison and his investors. As he was negotiating British rights for

J. P. Morgan was one of the first to invest in the Edison Electric Light Company. *(Library of Congress)*

Edison's electric light, Lowrey wrote to Edison that he should have "money enough not only to set you up forever but to enable you at once if you desire to build and formally endow a working laboratory such as the world needs and has never seen. I should like immensely to see your name given to a place of that sort which while conducted as nearly as possible on self supporting principles would give a fair opportunity for indigent but ingenious men to have their ideas exploited or exploded as the case might be."

The money from the investors was used not only for the supplies for the experiments and to hire more men, but to build a brick building at Menlo Park to test generators, and a new, larger office and library.

Edison was so confident of success of his planned system that he announced to a reporter from the *New York Sun* "I have it now! With the process I have just discovered, I can produce a thousand, nay ten thousand—[lamps] from one machine. Indeed, the number may be said to be infinite."

But much work had to be done first. In addition to devising a long-lasting, practical bulb, Edison and his team would have to invent all the parts of an electrical transmitting system: switches, conduits, generators, and meters. As the search for the right ingredients for a light bulb and electric system continued, Edison added experts in chemistry, physics, glassblowing, and other specialties to his team.

Edison's method of operation can be seen in his search for the perfect lamp. He wrote to a friend,

> I have the right principle and am on the right track, but time, hard work, & some good luck are necessary, too. It has been just so in all my inventions. The first step is an intuition and comes in a burst. Then difficulties arise. This thing gives

A drawing of Edison and his team experimenting with light bulbs in 1879. *(Courtesy of Keystone/Getty Images)*

out then that. "Bugs" as such little faults and difficulties are called, show themselves. Months of intense watching study & labor are required before commercial success—or failure—is certainly reached.

Two days after their visit to Wallace's foundry, machinist Charles Batchelor conducted experiments on lamps with thermal regulators. The idea was that when a filament became so hot that it was about to melt or burn, the regulator would turn off the current. When the filament cooled off, the current would come back on. Edison and Batchelor thought this was the answer to improving the longevity of the bulb.

Eventually, they decided the regulators were too complicated and abandoned them. Also, the lamp turning off and back on proved annoying.

One consequence of using thermal regulators would be to place the lights in a parallel configuration. Arc lights were wired in series and when one went out, breaking the current, the rest did as well. For parallel circuits, wires branch off the main one to light each lamp, so one can be turned off or burn out and not affect the others.

Edison had realized something that other inventors had not seen. A light bulb needs a filament with high resistance, which seems contrary to logic. Resistance is opposition to the flow of current. The amount of resistance is related to the size, length, and components of a wire. A short thick wire made from a good conductor will show low resistance: current will flow easily. High resistance comes from a thin, long wire of a poor conducting material. One way to obtain high resistance was to lengthen the wire, so they planned the filaments to be tightly wound spirals.

The measurement of electrical resistance came from Georg Ohm. In 1825, he stated Ohm's Law which states that current is equal to voltage divided by resistance.

$$I = V/R$$

One way to understand these terms is to think of water flowing from a reservoir into a pipe. If you increase the pressure on the water in the reservoir, more water flows out of the pipe. You have increased the current. If you use a bigger pipe, more water will flow and in electrical terms, you have lower resistance. In reverse, if the pipe is smaller, resistance is higher.

Edison in the late 1870s *(Library of Congress)*

Edison saw that each light would need a tiny part of the current to operate. Low resistance lights would require a huge amount of current and a thick copper wire to produce enough current to light the bulbs at the end. The heavy wire would be expensive due to the costs of copper. A wire with high resistance would conduct the electricity needed to light many bulbs with a lower cost.

Another electrical parameter is Joule's Law. James Prescott Joule had studied electricity and formulated his law only a few years before Edison was born. The law relates the amount of heat given off when a current is generated. The mathematical formula is:

$$H = RI^2$$

The formula shows that the heat produced is related to both the resistance and to the current. Edison knew he needed to generate as little heat as possible to keep the filaments from melting. The resistance is important, but the fact that the amount of current is squared means that is a more critical component. A higher resistance would allow a lower current and still keep the heat at a lower level.

To mitigate the problem of filaments burning up, Edison decided to remove the gas from the glass bulbs. At first, he was working with platinum for the filament. When it was heated, gases escaped, causing the wire to break. Those gases would not be present in a vacuum, so the filaments would heat up enough to glow. So he bought the best vacuum pumps that he could find and even made improvements to them.

As the search for the perfect filament material continued, Edison worked to find the right power source for his plan. Edison's previous experiments with electric lights had been powered by batteries. For a large system he needed the power of a generator. He bought one manufactured by Wallace and Moses Farmer. The dynamo could light eight arc lamps at once and Edison saw the capability to transmit power with such a generator. He also bought other dynamos and ordered some of his team to compare them.

After testing, Edison decided that the generators he had bought were not what was needed. He would have to build his own, working to develop "the machine that will give the greatest amount of electricity per horse power."

For the generator invention, he hired Francis R. Upton, who had received the first master's of science degree in physics at Princeton University. Upton had also studied in Berlin, Germany. Upton came up with a design for a new generator. Two four-and-a-half foot upright iron poles were the magnet. A hollow cylinder wrapped with copper wire became the armature as it rotated between the upright pieces. A steam engine provided the rotation power. The men nicknamed the odd-shaped machine the long-legged Mary Ann. The Mary Ann was an outstanding example of the electromagnetic machines envisioned by Michael Faraday, Edison's hero.

The efficiency rating of most generators at the time was only 60 percent but the Mary Ann reached the unheard of level of 80 percent. Upton saw the value of this advancement, "we have now the best generator of electricity ever made and this in itself will make a business."

Another obstacle for making a light bulb was forming the glass shape. Edison hired a recent immigrant from Germany, Ludwig Boehm, for the job. Boehm was an experienced glassblower.

Experiments to find a good filament continued. Edison tried platinum and a platinum-iridium alloy. The platinum filaments worked better than other materials, but there were two problems: the metals were very expensive and the resistance was too low. Edison wrote to mining experts all over the country to see if a cheaper source could be found. The search was fruitless, so he turned to other substances.

A drawing of Edison's high-efficiency generator *(Library of Congress)*

The slow progress with the lighting system annoyed the investors. They appealed to Grosvenor Lowrey, Edison's attorney and front man with the investors. He told Edison he needed to demonstrate some part of his experiments before investors pulled their money out of the project.

So on the evening of March 26, 1879, a group of the financiers boarded a train for Menlo Park. Edison greeted them warmly and escorted them to the finely furnished library. He explained the progress they had made. Then he took them to the laboratory for a demonstration. In the darkness of the machine shop, Edison turned on twelve incandescent bulbs, the best they had at the time. The visitors were impressed. A reporter from the *New York Herald* wrote, "The light given was clear, white and steady, pleasant to the eye." The discontented talk stopped.

In the middle of the research for a perfect light filament, Edison was continuing work on other projects. One of Edison's projects entertained his employees. He built an electric train that chugged around the property at Menlo Park. It ran on narrow gauge tracks for a distance of about one-third of a mile. The locomotive was a wheeled platform powered by a Mary Ann generator laid on its side. Power ran through the rails from stationary dynamos and to the generator which drove the wheels. The engineer's seat was a wooden crate and the train pulled passenger cars carrying up to twenty people. The train could get up to speeds of forty miles an hour. One time, Grosvenor Lowrey witnessed an accident with the train. John Kruesi was driving it when "the train jumped the track . . . throwing Kruesi in the dirt, and another man in a comical somersault through some underbrush." No one was seriously hurt, but Kruesi got up with his face bleeding and

Edison at the throttle of his electric train. *(Courtesy of AP Images)*

as Lowrey described, "I shall never forget the expression of voice and face in which he said, with some foreign accent: 'Oh! Yes, pairfeckly safe!'"

Edison had the cab of the locomotive closed in after that, and later built a larger locomotive and conducted experiments with it. But he never developed the electric train past that point.

Meanwhile, members of his team, including his nephew Charley, devised improvements to the telephone. Charley showed signs of becoming an outstanding inventor like his uncle. In spring 1879, Charley and others demonstrated the telephone to the British Royal Society and the Prince and Princess of Wales. The presentation convinced the British to accept Edison's phone system, and they formed the Edison Telephone Company of London.

Charley traveled to Paris, but in October 1879, Edison received word that his nephew was sick and not expected to live. All work at Menlo Park stopped and on October 19, a telegram arrived with the news that Charley had died. Edison arranged for Charley's body to be brought back to Port Huron.

In the middle of this time of grief, Edison found the right material for the filament of his lamp. He discarded metals because of their low resistance and high cost and returned to carbon. Several materials were carbonized—baked until

they charred—and tested in the bulb. They tried paper, cardboard, and lampblack. Then on October 22, Batchelor carbonized a piece of cotton thread. The charred thread was carefully looped into horseshoe shape and attached into a glass globe. After a vacuum was created, the current was turned on. The bulb shone brightly at a rating of thirty candles and it lasted almost fifteen hours. The resistance of the carbon met Edison's calculations.

Edison applied for a patent on the carbon filament in early

An Edison light bulb (*Courtesy of AP Images/Mike Derer*)

Light Bulbs: Then and Now

Today's light bulbs are very similar in design to Edison's, but there are some differences. Inside the glass bulb is a tiny filament, now made of tungsten metal. Surrounding the filament is not a vacuum, but a gas such as argon is introduced into the bulb. The argon prevents evaporation of the tungsten atoms as they are heated by electricity and begin to glow. The temperature in an incandescent bulb reaches about 4,000° F or 2,200° C.

November. They found that carbonized cardboard made a better filament than cotton thread.

Edison's financiers were pushing for a public demonstration of his light system, so Edison prepared multiple lights and electric wires to be strung around Menlo Park. A reporter wrote an article about the event, headlined "The Great Inventor's Triumph in Electric Illumination." People began to journey to Menlo Park long before the planned demonstration date of December 31. They wanted to see the most recent invention by the Wizard.

People traveled by wagon, horseback, and train to arrive at the laboratory at Menlo Park. The *New York Herald* printed that "Menlo Park [was] thronged with visitors coming from all directions to see the 'wonderful' electric light." All of these visitors arrived on December 30, the day before the switches would be turned on. They tromped through the laboratory, and Edison and his team answered their questions as best they could. In the midst of the crowds, the preparation for the following night continued.

The reporter from the *Herald* wrote about the festivities on December 31, 1879:

> Extra trains were run from east and west, and notwithstanding the stormy weather, hundreds of persons availed themselves of the privilege. The laboratory was brilliantly illuminated with twenty-five lamps, the office and counting room with eight, and twenty others were distributed in the street leading to the depot and in some of the adjoining houses. The entire system was explained in detail by Edison and his assistants, and the light was subjected to a variety of tests.

Edison and the others gave several demonstrations. They stuck one lit bulb in a jar of water and it kept glowing for four hours. They switched lights off and on multiple times to simulate the wear and tear of household use. Another demonstration was an electric motor running a sewing machine and a water pump. One man wanted to know "how you got the red-hot hairpin into that bottle?"

As the hour turned to a new year and a new decade, people were astonished by the inventions they saw at Menlo Park. They were the first to enjoy the electric age.

Edison's work was just beginning, though. His original plan to illuminate part of Manhattan still had much to overcome.

Six
Lighting Up Manhattan

One of the visitors to Menlo Park on New Year's Eve was Henry Villard. He was one of the investors in the Edison Electric Light Company, but he came back later to see Edison about a steamship that he was building, the *Columbia*. He wanted to install Edison's generators and light system on the *Columbia*, which was being outfitted in New York harbor. Edison took on the job as an opportunity for good publicity.

To install the lighting system, Edison and his team had to design the first lamp socket, key switch, and fuse. A fuse works to prevent wires from overheating and starting a fire. When too much current flows through the fuse, the center connector breaks before damage can be done to the rest of the system. Four Mary Ann generators powered the system on the ship and 150 lights were installed.

The system on the *Columbia* was tested successfully on April 28. On May 9, the steamship left New York and sailed around South America and then north in the Pacific to Portland, Oregon. At every port where it stopped, people gazed in awe at the steady bright illumination coming from its Edison lights.

Edison knew from the success of the system on the *Columbia* that he could market his power systems to isolated customers who could not be attached to generators by long wires.

Wealthy banker J. P. Morgan requested a stand-alone system for his mansion in New York City. A power plant was built in a cellar beneath his stables and an underground conduit transported the wires into the house. An engineer reported for duty to tend the power plant every afternoon at 3:00 o'clock. His first chore was to fire up the boilers so that power would be available by 4:00. He stayed until 11:00. Sometimes his shut off of the power plant darkened the house in the middle of entertaining.

After a year, Morgan requested Edison to upgrade his system as he always liked to be the first to have anything new. He said, "I hope that the Edison Company appreciates the value of my house as an experimental station." After the upgrade, he hosted a party for four hundred guests to show off his lighting system. Some of the guests ordered their own systems, and Morgan paid for systems to be added to St. George's Church and to a private school.

For every part of the Manhattan lighting system, something new would have to be invented. Overcoming these obstacles took all the ingenuity of Edison and his men.

New York had been chosen for Edison's first urban system because most of his investors lived there, and there would be a good concentration of customers. Through 1880, work on all the components of the system continued.

By this point, Edison employed about sixty-five men and that number would increase through the lighting project. To make the glass globes for the bulbs, Edison hired Ludwig Boehm. Trained as a glassblower, Boehm constructed the first bulbs. When mass production was needed, Edison turned to Corning Glass Works.

Edison in 1880 *(Library of Congress)*

One challenge was the clamp that held the filament in the bulb and connected it to the electrical wires. The original plan was to use platinum, but it was expensive. The next choice was nickel, but that metal proved hard to shape. One of Edison's employees devised a way to attach the filament without clamps by plating the filament to the wires with copper.

In July of 1880, Edison was still searching for an even better filament material than cotton thread. He had made a survey of natural fibers, saying, "Before I got through I tested no few than 6,000 vegetable growths, and ransacked the world." As he cooled himself with a handheld fan one hot day, he noticed that the fan was made of bamboo. He cut a strip off of the fan and gave it to Batchelor to be tested. It worked well, but now the problem was finding a good supply of bamboo. Edison requested that people traveling to the areas collect samples from all over the Pacific nations, and South and Central America. The chosen species came from Japan, and bamboo was used for the filament for the next ten years.

Designing a lamp socket was also an obstacle. The first ones were wooden stems with copper strips to make contact. This design would work fine for upright lamps, but not for ceiling lights. After many designs were tested, the one chosen was a screw mounting made of plaster of Paris with metallic contact points.

The components of the light bulb proved the most challenging both to invent and to manufacture. By 1884, Upton wrote that 2,774 experiments on the lamp had been performed at an estimated cost of more than $70,000. Still Edison persisted, sure that the next experiment would prove to be the crucial one.

The Edison Effect

During experiments on the electric light bulb, Edison came across another pure science curiosity. He noticed that a blackish deposit would form on the inside of the bulbs. His theory was that carbon particles were becoming electrically charged and migrating to the glass surface. But how could this happen in a vacuum? In tests in 1883, Edison put a tiny platinum plate inside one of his bulbs. When a meter was attached to the plate, it showed that an electric current flowed between the filament and the plate. Edison wondered how that was possible without a wire. He patented a special lamp showing the strange current and he went on to other experiments. The phenomenon became known as the Edison effect or thermionic emission.

At the time of Edison, no one knew how this phenomenon arose. Now we know that it occurs because electrons can move across empty space. The Edison effect is a foundation principal for all electronics.

Edison's experience with telegraph wires led him to prefer underground installation of the power cables. Wires strung overhead are susceptible to winds and ice storms. Besides, all the telegraph companies working in New York had already strung wires in a spaghetti-like mess over most of the main streets. When a wire broke, it was hard to know what company it belonged to.

Trenches were dug around the laboratory to test the idea. Insulation of the wires proved to be the biggest hurdle. After several tries, they settled on wrapping the wires in muslin, and coating that with a mixture of paraffin, asphalt, tar, and linseed oil.

An 1882 drawing of workers installing underground electrical wiring in New York City. *(Library of Congress)*

In December 1880, the famous French actress Sarah Bernhardt came to Menlo Park, wanting to meet "*le grande* Edison." She boarded a train after one of her performances in New York and arrived at Menlo Park at about 2:00 am. Edison gave her the tour, assigning one man to watch over her long swishing dress to make sure it didn't get caught in the equipment. He recorded her voice with the phonograph and flicked the outdoor lights on and off for her amusement. When she left she exclaimed, "*C'est grand, c'est magnifique!*" ("it is large, it is splendid!")

Famous French actress
Sarah Bernhardt was
given a personal tour of
Menlo Park in 1880. *(Library
of Congress)*

Before construction
of the system could start
in Manhattan, Edison
would need the approval
of the city officials.
In December 1880,
the members of the
New York City Board
of Aldermen came to
Menlo Park to view
Edison's work. Edison
provided an elegant
meal complete with
champagne, then dem-
onstrated four hundred
lamps lighting up the evening gloom. The display persuaded
the aldermen; they gave Edison a franchise to construct his
system in their city.

Meanwhile, Edison's competitors in the electric lighting
field were as hard at work as he was. The same night that the
New York aldermen visited him at Menlo Park, the Brush
Electric Company lit up Broadway with an arc lighting system. The *New York Evening Post* recorded that the lights
shone "with a clear, sharp, bluish light resembling intense

moonlight." From these brilliant lights, Broadway obtained its nickname as "The Great White Way."

The area that Edison would set alight was chosen: ten blocks bordered by the East River, Nassau Street, Spruce Street, and Wall Street. Edison bought two buildings on Pearl Street, near the center of the square, for his generators. The Edison operations began moving from Menlo Park into New York City. The machine shop relocated to 104 Goerck Street. The Edison Electric Tube Company, responsible for the underground installation and equipment, settled in at 65 Washington Street. In Menlo Park, Upton remained as manager of the lamp factory, though later that factory moved to East Newark, New Jersey.

As Edison said later,

> The Pearl Street Station was the biggest and most responsible thing I had ever undertaken. It was a gigantic problem, with many ramifications . . . All our apparatus, devices and parts were home-devised and home-made. Our men were completely new and without central station experience. What might happen on turning a big current into the conductors under the streets of New York no one could say.

Edison moved his headquarters into a mansion at 65 Fifth Avenue. He remembered another trip to New York, saying, "We're up in the world now. I remember ten years ago—I had just come from Boston—I had to walk the streets of New York all night because I hadn't the price of a bed. And now think of it! I'm to occupy a whole house in Fifth Avenue!" He had the mansion wired for lights and they blazed out every evening. On the walls of his office, he attached maps showing the area he intended to serve with his electrical system, called the First District.

With Edison's move to Fifth Avenue, the era of Menlo Park was ending. His inventions made there included the phonograph and the electric light. He kept his house there for a while, but from then on, Edison would live and work in other places. But his invention factory set the standard for research laboratories around the world.

Edison's right hand man, Charles Batchelor, had sailed to England and France to exhibit the lighting system in London and Paris and to manage Edison's operations in Europe. Shipped along with him was a new, enormous generator like the one that Edison was using for his project in New York. The dynamo weighed thirty tons and was nicknamed Jumbo. Edison's display of the lighting system at the International Electrical Exhibition in Paris was highly praised. A London

This drawing depicts the different levels of Edison's New York power station. *(Courtesy of Mary Evans Picture Library/Alamy)*

newspaper proclaimed, "Mr. Edison's exhibition is the wonder of the show."

An even bigger Jumbo was on its way to England to be the generator for a system lighting Holborn Viaduct in London. Nearly a thousand lamps lit up shops, streets, restaurants, and hotels. Edison's electrical systems were delighting Europeans as much as Americans.

The generators at the Pearl central station were ready for a test. In July, two Jumbo dynamos were connected and fired up by two steam engines. Edison remembered, "At first everything worked all right. Then we started another engine and threw them in parallel. Of all the circuses since Adam was born, we had the worst then! One engine would stop and the other would run up to a thousand revolutions; and then they would seesaw." The building shook and sparks flew. Edison and another man shut down the engines. By inspection, they found a defect in the steam engines and Edison switched to another kind before the system had its true opening.

Testing of the components of the electrical system continued and proved successful. So on September 4, 1882, the project was ready for operation. At 3:00 p.m., Edison threw the switch and 33,000 lamps lit up in buildings including the Drexel-Morgan Building and the *New York Times.* The reporters at the *Times,* "men who have battered their eyes sufficiently by years of night work to know the good and bad points of a lamp," complimented the quality of the light compared to gas lights. The electricity flowed through almost 100,000 feet of cable, costing more than $180,000 to install.

The *New York Herald* had an onsite power plant, supplying electricity for its lights. The newspaper reported on Edison's Pearl Street station that "the lighting . . . was eminently satisfactory . . . the luminous horseshoes did their work well." Edison summed up his feelings about the event: "I have accomplished all I promised."

Direct Current versus Alternating Current

By the time Edison finished the Pearl Street Station and lit up Manhattan, he was embroiled in a battle over the preferred type of electricity. All of Edison's systems worked by direct current, the kind produced by a battery. Direct current or DC flows in one direction. But alternating current, AC, is the electricity produced by a dynamo. The current cycles back and forth. To get direct current from a dynamo, a commutator must be used to switch the AC into DC.

There are several advantages to AC. Commutators are usually brushes and they wear out quickly, so using AC as produced from a dynamo would be less maintenance. Also, a DC power station could only provide power up to a half of a mile away. AC can be transmitted over long distances, with little loss of power.

The biggest advantage to DC is that it runs on a lower voltage than AC. Alternating current is transmitted at a very high voltage and stepped down in voltage by transformers before it enters homes or businesses. People can be electrocuted by AC power. At the time, another benefit was that motors had been designed to run on DC and not on AC.

A Serbian immigrant came into the picture at this time. Nikola Tesla had worked for Edison for a while, but then left to work on his plans for AC. Before coming to the United States, he had envisioned an AC motor, power system, and all the parts.

Another proponent of the AC system was George Westinghouse. An inventor in his own right, Westinghouse had designed an innovative air brake for the railroads. Now, he was entering the electrical industry with big plans; to build a power plant at Niagara Falls to provide electricity for a large region, including Buffalo, New York.

George Westinghouse *(Library of Congress)*

Edison's ego would not let him see that AC was the electricity of the future. He fought against Westinghouse's plans, saying, "Just as certain as death, Westinghouse will kill a customer within six months after he puts in a system of any size."

Westinghouse persisted and his system at Niagara Falls was built. Tesla's AC motor became the industry standard as did alternating current. Edison lost the battle of the currents. Today our electricity is alternating current, measured at 60 cycles per second.

seven
New Industries, New Ideas

Edison's attention returned to family affairs. Mary had seemed more content when they first moved to New York City, but she still resented the time that her husband spent with his work, and her health had never recovered after the birth of their third child, William Leslie, in 1878. In January 1882, her health worsened, leading to a nervous breakdown, and Edison took off time in February and March to travel with her to Florida. As her daughter remembered, "Mother was never happier than when we went to Florida, for then Father belonged entirely to her."

After return to the home in New York, Mary became so weak that her doctor told her to give up housekeeping chores. Even with servants, they couldn't keep up the house, so the family moved into the Clarendon Hotel. In 1883, they took a winter trip to Florida again.

Soon after their return, Mary's father died. She grieved for him and then seemed to improve. But by summer of 1884, she was very ill, complaining of headaches and weakness. The Edisons returned to the home at Menlo Park and unexpectedly, on August 9, 1884, Mary died. She was only twenty-nine, and probably died from a brain tumor. Marion remembered that she found her father "shaking with grief, weeping and sobbing so he could hardly tell me that mother had died in the night."

Mary's mother took the two sons, Thomas and William to live with her. Marion stayed with her father and they became very close. Notations that appear to be her homework often appeared in his lab notebooks.

After his wife's death, Edison rarely returned to Menlo Park. The laboratory fell into ruin. Already prone to immersing himself into his work, Edison focused on his New York lab, jumping from project to project. He worked on improvements for the telegraph, the telephone, and the phonograph. His telephone work brought him into contact with Ezra Gilliland, who worked for American Bell Telephone, and together they made many improvements to the telephone.

At the World Industrial and Cotton Centennial Exposition in New Orleans, Edison met eighteen-year-old Mina Miller. He fell in love with her almost immediately. Her father, also a sometimes inventor, was pleased that his daughter was drawing Edison's attention.

That summer, Edison traveled to Chautauqua, New York, where the Millers had a summer home. Mina's father, Lewis Miller, had been one of the founders of the Chautauqua Institute, which provided lectures, concerts, and other entertainment each summer. Miller was a strong supporter

Mina Miller *(Library of Congress)*

of education for men and women, including college. His daughters attended college preparatory schools and two of them went to Wellesley. Mina attended a school in Boston and studied music. Therefore, when Edison visited his friend Gilliland in Boston, he could also see Mina.

To communicate privately, Edison taught Mina Morse code and they would tap out messages to each other and no one else would know. Edison even proposed this way and to his delight, Mina answered, "dash dot dash dash dot dot dot dot [Yes.]"

Edison and Mina were married on February 24, 1886, and traveled to Fort Myers, Florida for their honeymoon. Edison had bought property there for a winter home.

Before the wedding, Thomas and Mina had settled on their new home in the New York area. Mina preferred a country home to one in the city. So Thomas bought a mansion called Glenmont in West Orange, New Jersey for $250,000. The home was huge, with twenty-nine rooms.

Edison's three children moved in with the newlyweds and Mina experienced some tough times with daughter Marion. Dot, as she was called, resented another mother and someone taking over the time that she had spent with her father. Edison was never close to his oldest sons. His neglect of them in their childhood caused a barrier that was never completely overcome.

Edison's Glenmont mansion in West Orange, New Jersey *(Library of Congress)*

Edison was building a new laboratory complex in West Orange—a much larger one than he had at Menlo Park. The laboratory building was three stories and 40,000 square feet, comprised of lab rooms, a library, offices, machine shop, and a power plant. Other buildings were constructed for chemical and metallurgic labs, electrical testing, and woodworking.

The best equipment and machines were bought for Edison's new laboratory. He filled the library with scientific books and magazines, and the workshops with quality materials and tools. In his view, "the most important part of an experimental laboratory is a big scrap heap."

Through the years, Edison's West Orange facility expanded several times. By about 1915, it covered more than twenty acres and employed 10,000 people.

One of the workshops at Edison's West Orange laboratory complex. (*Library of Congress*)

Edison had little to do with the management of his companies. In 1889, the investors of several of his companies merged into one called Edison General Electric Company. By 1892, the company was called General Electric, taking Edison's name out. Edison had been hurt by the decision to remove his name from General Electric Company (GE). But he didn't quit. He said, "I'm going to do something now so different and so much bigger than anything I've ever done before people will forget that my name was ever connected with anything electrical."

His new project was in the field of mining and refining. When searching for the perfect filament for his light bulb, Edison had investigated the mining of platinum. On his trip west, he had toured gold and silver mines. Now he decided to try to develop a new way to mine iron. Using money from GE stock that he sold, he bought some iron ore mines and built a plant to try to separate iron from ore by using electromagnets. The ore in the area was a low grade, meaning low iron content. But Edison was convinced that his method would separate the iron efficiently enough to be profitable. It didn't turn out the way he wanted. In 1890, rich iron ore deposits were found in Minnesota, putting Edison's plant out of business. Many called this venture "Edison's Folly." When he ended the operation, Edison had spent all the money he received from his stock in General Electric. He said, "Well, it's all gone, but we had a hell of a good time spending it!"

He did develop one innovative idea at the iron ore plant. He used a system of conveyor belts to move the ore from one place to another. This system of conveyors was implemented in Henry Ford's first automobile factory and is still in use in mass production of most factories today.

Returning to his favorite invention, Edison made improvements to the phonograph. Alexander Graham Bell, his cousin, Chichester Bell, and Charles Tainter had gotten patents when Edison's expired. They called their machine the gramophone. To compete, Edison improved his phonograph, including building one powered by an electric motor.

Jesse Lippincott bought both the Bells' company and Edison's. Each inventor manufactured his version of the talking machine and Lippincott handled marketing.

Edison also designed a talking doll. When a child turned a crank, the miniature phonograph inside the doll would recite nursery rhymes. The doll was a failure because the crank would break and the needle did not stay in the groove well enough.

In 1889, Edison and Mina traveled to Europe. They were greeted in Paris by the president of France and when they attended the Paris Opera, the orchestra played "The Star Spangled Banner."

Edison met Alexander Gustave Eiffel, designer of the Eiffel Tower, and remarked later that Eiffel was friendly and humble, and one of his favorite people in France. He also wrote in Eiffel's guest book, complimenting the French engineer on his remarkable and original achievement. The Edisons also visited the scientist Louis Pasteur at his laboratory.

After France they traveled to Berlin and visited with Hermann von Helmholtz, a pioneer in the study of sound waves. After stops in Belgium and England, they returned home.

Another idea was catching Edison's eye. Englishman Eadweard Muybridge visited Edison at the West Orange lab

For Vacation Fun and Music You Need an

IMPROVED EDISON PHONOGRAPH

THE Phonograph solves the problem of music and entertainment in the summer home or camp. Don't fail to make one a member of your vacation party.

No matter where you go, you can transport a veritable theatre with you. Around the camp-fire, on the launch, or at the farm, the Phonograph is ever ready to entertain you with the world's best music. Rainy days yield hours of pleasure.

Evenings can be spent listening to whatever kind of vocal or instrumental music suits your fancy, or the Phonograph will provide music for a two-step on the veranda or a reel on Nature's carpet.

NEW SERIES OF GRAND OPERA RECORDS

The success of the first series of Edison Grand Opera Records surpassed our most sanguine expectations. The second series of ten is fully equal to if not better than the first. They consist of favorite selections from standard grand opera rendered in French, German, and Italian, by Constantino, Knote, van Rooy, Scotti, Berti, Dippel, Resky, and Signora Resky. Now on sale at all dealers.

Hear the Edison Phonograph at the dealer's free of charge. Write for Booklet "Home Entertainments With the Edison Phonograph," and name of nearest dealer.

National Phonograph Co., 18 Lakeside Ave., Orange, N. J.
New York: 31 Union Square   Chicago: 304 Wabash Avenue

The advertisements in Everybody's Magazine are indexed. Turn to page 3.

This ad indicates how quickly Edison's phonograph became a popular household item.

During an 1889 trip to Europe, Edison and his wife were able to meet Alexandre Gustave Eiffel, the designer of the Eiffel tower. *(Library of Congress)*

complex. He showed Edison his groundbreaking photographs. He placed multiple cameras along a racetrack. The cameras were triggered to take a picture as a horse ran by, providing multiple images. One question that had been debated for centuries was answered by his pictures. A horse does have all four legs off the ground during a gallop.

Muybridge demonstrated his pictures in a zoetrope, a rotating cylinder which gave the appearance of movement.

Eadweard Muybridge *(Library of Congress)*

Edison was intrigued and began to work on moving pictures, saying he wanted to "do for the eye what the phonograph does for the ear."

Several inventors had already made advances that helped Edison. George Eastman had made a new film, long strips of flexible celluloid. During a trip to Paris, Edison had been shown film with a perforated edge that could be used to advance the film.

Edison's employee, W. K. L. Dickson, was put in charge of photography at the lab. Putting these two inventions together, they created a film that could be fed through sprockets and

One of the famous series of photos Muybridge took of a running horse. (*Library of Congress*)

a lens and shutter system to quickly expose each frame of film. Powered by an electric motor, the machine was called a kinetograph, meaning motion recorder. To show the movie, another machine, the kinetoscope, was invented. The kinetoscope was a cabinet with a peephole on top. A person would peer through a magnifying lens as the motor and sprockets moved the film to display the pictures. People paid five cents to watch a few minutes of film.

One of the first films was a ten-round boxing match. To see the entire film, a person had to move to several kinetoscopes in sequence. The film was a big hit.

In 1893, an odd building appeared on the grounds of the West Orange complex. Called Black Maria, it was a wooden building covered inside and out with black tar paper. This

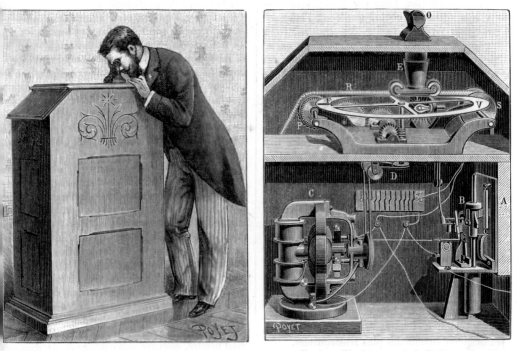

Fig. 2. — Vue extérieure du Kinétoscope.   Fig. 3. — Mécanisme moteur du Kinétoscope.

Exterior and interior drawings of Edison's kinetoscope. *(Courtesy of Mary Evans Picture Library/Alamy)*

was the first motion picture studio. A hole in the roof allowed sunlight in for the best photography and the whole building could be turned to follow the sun through the day.

The films shot in the Black Maria were silent. Edison worked hard to match up the sound of a phonograph with the motion picture but it was too complex a process. The studio made a number of silent films, including several from five to fifteen minutes long about children waiting for Santa, a fairy kingdom of glaciers, and the Monarch of Arctic (a polar bear). There are several recordings of Edison and his team. Edison is seen cupping his right hand around his ear to enhance his poor hearing.

Edison sitting outside of his office in 1895. *(Library of Congress)*

In 1903, Edison's studio produced *The Great Train Robbery*, one of the first narrative movies. In it, outlaws rob a train in the Wild West and are chased for their crimes. Directed and conceived by one of Edison's cameramen, Edwin S. Porter, the film utilized many groundbreaking techniques, including cross-cut editing and location shooting. In the final shot, one of the outlaws points his gun right at the audience, causing excitement and fear among early viewers.

Eventually, Edison's motion picture business merged with two others and became the Motion Pictures Patents Company. The company produced, marketed, and distributed almost all movies in the United States for the next ten years.

While Thomas was producing new inventions at the West Orange facility, Mina was having children. Like his first wife, Mina had three children; a girl and two boys. Madeleine was born in 1888, Charles in 1890, and Theodore in 1898. Theodore was named for Mina's brother who died in the Spanish-American War. Edison spent more time with these children than he had with his first. Still he worked long hours, sometimes all night.

As the twentieth century approached, Edison's stake in the phonographic business had grown rapidly. When Lippincott died, Edison got back the distribution rights to his phonographs and formed a new company, National Phonograph Company. He replaced the difficult electric motor with a wind-up version. By 1900, sales had grown to $250,000 a year.

He developed a method to mass produce recorded cylinders and began selling them. The recording industry was divided into three companies; Edison's National Phonograph Company, Columbia Phonograph Company, and the Victor Talking Machine Company. Columbia and Victor were both selling flat records which had a longer recording time; however, Edison's cylinders had better sound quality. In 1909, Edison began producing flat records. The industry had chosen that medium as the standard.

Edison had experimented with designing a helicopter in 1885. He realized that whatever flying machine was built, it would be heavier than air and would need some way to give it lift. He was thrilled when in 1903 the Wright brothers flew their plane.

In 1910, Edison's companies were combined into Thomas A. Edison, Incorporated. Edison was named president and chairman, but he was sixty-four years old, so much of the daily work was left to his employees.

After his ore milling operation had closed, Edison turned to another mineral product. He became interested in cement and concrete. Cement had been invented by the Romans; an Englishman, Canvas White, revived the material in the 1800s and discovered a method to heat the rocks and grind them.

Edison had already found that the most profitable product from the ore milling operation was the waste sand from the crushing process. He sold sand to railroads, building contractors, and asphalt companies. Then he began selling some to cement companies for use in buildings. His interest piqued, he began investigating cement and concrete.

Edison learned that "the quality of a cement depends upon two things. 1st: The proper proportioning of the ingredients and burning. 2nd to fine grinding." He had the best crushing and grinding equipment in the world already from the failed ore mill. So he went to work to perfect the first quality.

He built a rotating kiln that roasted the cement rock and limestone into what is called clinker. He experimented with different designs, fuels, and lengths. He also experimented with the chemistry of the ingredients of cement. Because different rocks contain different concentrations of minerals, each batch had to be tested regularly to ensure the quality and consistency of the concrete.

Edison also tinkered with building concrete homes. A house made of concrete does not burn, is impervious to insects such as termites, and is long lasting. Molds for houses were built and once the molds were ready, an entire house could be poured in about six hours at a cost of $1,200. However, constructing the molds was very expensive, so the idea of concrete homes did not catch on until after Edison's time.

Edison's cement company closed during the Great Depression, but it made a strong impact first. Edison's concrete was used to build Yankee Stadium in New York City and part of the Panama Canal. Other manufacturers of cement accepted his technology for the industry.

The 1925 opening ceremony at Yankee Stadium *(Library of Congress)*

eight
World War I and Beyond

W hen Wilhelm Röntgen discovered X-rays in 1895, Edison was immediately fascinated. He telegraphed to a friend and colleague, "How would you like to come over and experiment on Rotgons [Röntgen] new radiations. I have glassblower and Pumps running and all Photographic apparatus. We could do a lot before others get their second wind."

Edison tested more than 150 designs of vacuum tubes for X-rays in two months and also tested power sources, time to expose photograph plates, and how much X-rays penetrated different materials. Unfortunately at the time, no one realized the danger of using the radiation from X-rays without proper protection. One of Edison's assistants later had to have his arms amputated because of damage from X-rays.

Edison and his team discovered that calcium tungstate provided the best image from X-rays and allowed doctors to

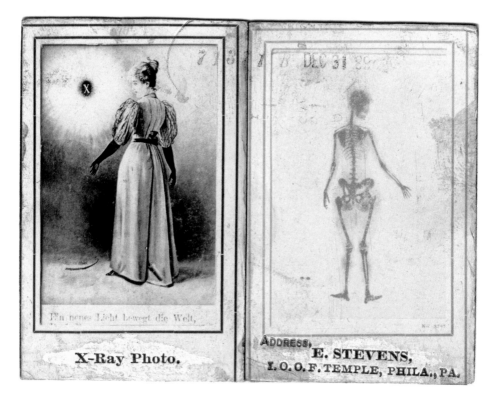

X-Ray Photo.

ADDRESS,
E. STEVENS,
I.O.O.F. TEMPLE, PHILA., PA.

An x-ray of a woman taken in 1896. *(Library of Congress)*

see inside human bodies. Using a calcium tungstate screen, Edison developed and produced the first commercially available fluoroscope, a device which allowed people to view inside a living body through X-rays. At the 1896 Electrical Exhibition in New York City, thousands of people stepped in front of an Edison fluoroscope and saw their own bones.

Edison and the President

In 1901, President William McKinley was shot and a bullet lodged near his spine. Doctors asked Edison to send a fluoroscope to find the bullet's location, but they thought that McKinley was recovering and didn't use the machine. The president died eight days later.

Edison became friends with another American icon, Henry Ford. The two met while Ford was working at Edison Illuminating Company of Detroit in the 1880s. Edison was told about Ford: "There's a young fellow who has made a gas car." The two discussed Ford's plans, and Edison encouraged him to keep at it. According to Ford, Edison, upon learning about Ford's gas engine, banged his fist down on the table. "Young man," he said, "that's the thing! You have it! Your car is self contained and carries its own power plant."

Years later, Ford, said in a newspaper interview,

> That bang on the table was worth worlds to me. No man up to then had given me any encouragement. I had hoped that

Henry Ford (left) talking to Edison in 1929. *(Courtesy of AP Images)*

I was headed right. Sometimes I knew that I was, sometimes I only wondered, but here, all at once and out of a clear sky, the greatest inventive genius in the world had given me complete approval. The man who knew most about electricity in the world had said that for the purpose, my gas motor was better than any electric motor could be.

Edison commented about Ford, "This fellow Ford is like the postage stamp. He sticks to one thing until he gets there." The compliment could apply to Edison as well.

Edison preferred an electric car to a gasoline one because it would run quieter and not produce fumes. But no one had been able to design a battery that would run an electric car for long distances. Edison decided, "I guess I'll have to make a battery."

Edison working in his lab in 1905. *(Library of Congress)*

Batteries were made of lead and filled with sulphuric acid. If you wanted more power, you needed a heavier battery. Edison wanted to design a lighter, stronger battery. More than 9,000 experiments later, he developed a new battery with plates of nickel, which is lighter than lead.

Once the batteries were in production, problems developed. They would not hold a charge as long as Edison hoped and leaks occurred. Edison stopped production and recalled the batteries sold. While he was working on the batteries, a visitor to the lab commented that it was a shame that Edison did so much work for no results. Edison answered, "Results! Why, man, I have gotten a lot of results! I know several thousand things that won't work!"

Edison's attempt to build a practical electric car was overtaken by the success of his friend, Henry Ford. By 1908, Ford was building enough inexpensive gasoline cars on his assembly line that electric cars were shoved to the side.

The friendship between Edison and Ford did lead to some different battery experiments. Ford wanted a way to start his Model Ts without using the hand crank. Edison's battery did not meet the test. It was unreliable in cold weather and did not retain power to crank the engine. Eventually, though, batteries did replace the hand crank.

Edison and Ford remained close friends. They took camping trips together, along with naturalist John Burroughs and tire manufacturer Harvey Firestone. The group explored the Great Smoky Mountains, the Adirondacks, the Everglades, and other parks. In later years, they were joined by President Warren Harding and still later by President Calvin Coolidge. Edison and Ford also were neighbors, having houses side by side in Fort Myers, Florida.

Edison relaxes during a camping trip in 1921. *(Library of Congress)*

In December 1914, much of Edison's West Orange complex went up in flames. An explosion in the film laboratory started the fire and the accumulation of various chemicals spread it quickly. As he watched the fire consume his factory buildings, Edison told his son, Charles, "Where's Mother? Get her over here, and her friends too. They'll never see a fire like this again." The main laboratory was saved and in a few days, Edison was rebuilding the rest helped by an interest-free loan from Ford. "I am sixty-seven; but I'm not too old to make a fresh start," Edison told Charles.

Edison's storage battery experiments had failed to start cars reliably, but they did impress the United States Navy. They were looking for a way to power submarines. As World War I broke out in Europe, the Navy was especially

This photo was taken the day after fire destroyed most of the West Orange laboratories. (*Library of Congress*)

interested in Edison's battery. They installed the batteries on their experimental submarine, the E2. However, the result was tragic. An explosion from hydrogen out of the batteries killed five sailors in 1916. The crew had not followed the safety regulations, but the incident was blamed on the batteries.

The Navy had named Edison chairman of the Naval Consulting Board in 1915. One of Edison's main tasks was to marshal technological advances to help in the war effort. He recommended that the Navy set up a research lab, but the Navy brass was not interested in a lab.

Still, Edison carried on with his work. He experimented with ways to detect enemy submarines, but the results were not successful, though he did develop a device to let submarine captains know when an enemy sub had fired a torpedo. He made many other innovations for the Navy as well: a sticky chemical that spread over the ocean to coat the lens of a periscope from an enemy sub; a loud-speaker telephone that could be used during battles; a glare eliminator for surface ships to allow sailors to see periscopes in the water; a coating to protect guns from seawater; and a fire-prevention chemical to help fight fires.

Edison (right) inspects the E2 submarine while serving as chairman of the Naval Consulting Board. *(Library of Congress)*

For his contributions, Edison was offered the Distinguished Service Medal by the Navy. However, he thought that others had made as great a contribution to the Navy Consulting Board as he had, so he refused to accept the award. Five years after the war ended in 1918, the Navy took Edison's advice and set up a Naval research lab in 1923.

President Woodrow Wilson honored Edison on his seventieth birthday in 1917. The president said, "I was an undergraduate at the university when his first inventions captured the imagination of the world. And ever since then I have retained the sense of magic, which what he did then created in my mind. He seems always to have been in the special confidence of Nature herself."

Edison tried to set up what he called "the clerkless shop." His idea was to have boxes with items for sale. A person would insert coins and choose the things they wanted. An electromagnet would open a chute, the item would drop down and the customer could reach it. Edison thought the system would be cheaper because there would be no salaries for clerks. Others thought his idea was crazy. Today vending machines, much like Edison's, are a common sight.

After World War I, the country's economy grew for a while. But when the Great Depression hit the country at the end of the 1920s, Edison was forced to cut his work force drastically, especially amongst his factory workers.

Edison was already unpopular with the men who worked in his factory. One worker said, "the people who worked very closely with Mr. Edison, as the years go by, have all very fond memories. However, its been my experience here in West Orange that every one of the folks who worked in the factory . . . I haven't found one person who expressed a good opinion of him . . . Even today you find around in West Orange people who say, 'Oh, he was rotten.'"

Edison believed that everyone should be willing to work as hard as he did. Even when he could see the damage that his constant work did to his family, he expected his employees to put in long hours. The wages he paid were at the low end of the scale. He also had no liking for labor unions. When machinists at the ore mill considered striking for extra pay for night and Sunday work, Edison broke the strike. He announced "all work on this plant is hereby suspended." After a couple of months, some of the workers were rehired, but others, especially those suspected of being the ringleaders of the strike, lost their jobs permanently.

As he grew older, work occupied Edison most of the time, but besides camping, his other hobby was fishing. Sometimes he would rent a boat and take friends on fishing trips in the waters near Long Island. Edison was willing to fish for hours, even if he got no bites. He also built a long dock out into the Caloosahatchee River near his winter home. The catch there was usually catfish, but as his wife said, "his greatest joy was catching a shark, which he would do occasionally." His sons also fished with him. Son Charles said, "[he] always used to say that fishing off the end of the dock was the greatest relaxation he knew of."

Edison and his family spent each winter at their home in Florida. They named it Seminole Lodge, in honor of the Seminole Indians who had once lived in the area. Edison had a laboratory supplied in the home, so he could continue experiments in the winter months.

In 1914, Edison, Ford, and Firestone were visiting in San Francisco. They saw an exhibit of what was called "Electric

Dinner." A meal was cooked on a stove powered by electricity, a revolutionary advance at the time.

While in California, they also visited the famous naturalist Luther Burbank. Edison was impressed with Burbank's plant experiments. "He is constantly busy creating new flowers and plants," he said. "This fellow knows that plants are not stagnant but can be changed to come up with new varieties of flowers with beautiful colors and forms."

The last Edison-Ford-Firestone camping trip took place in 1924. The group traveled to Vermont and stopped to visit President Calvin Coolidge. A story went around about their trip to a factory.

> Something went wrong with the car and they stopped near a farmhouse. The farmer . . . offered his help and at the same time started to lift the hood when Mr. Ford stopped him and said: "There's nothing the matter with that engine: I'm Henry Ford and I know all about engines." The farmer then suggested the trouble might be in the battery and Mr. Edison spoke up and said: "No, I'm Thomas A. Edison and I know all about batteries. That one is all right." The farmer began to look incredulous but tried again by suggesting the tires needed air and offered to pump them up, but Mr. Firestone put in with, "No, I'm Harvey Firestone and I made those tires; they're just right." The farmer exploded at this with, "Well, Ford, Edison and Firestone, eh? I reckon that little runt in the back seat's Calvin Coolidge?"

The year 1929 was the fiftieth anniversary of the light bulb. Henry Ford had a personal project that provided an ideal means for celebrating the occasion. Ford had decided to preserve important buildings and scenes of bygone days. Near his car plant in Detroit, he established Greenfield Village and built models of buildings, such as Independence Hall, or collected and rebuilt historic buildings celebrating accomplishments of people like songwriter Stephen Foster, Luther Burbank, and the Wright Brothers. He had Edison's Menlo Park laboratory moved to Greenfield Village and rebuilt just as it had been when Edison created the light bulb. He even reconstructed the boardinghouse where several of Edison's employees had lived.

On October 21, 1929, the Light's Golden Jubilee was celebrated at Greenfield Village. Ford escorted Edison through the Menlo Park lab reconstruction. When Ford asked him how it looked, Edison replied, "Well, you've got it about ninety-nine and one-half percent right." Ford asked, "What's the matter with the other one-half percent?" Edison answered, "We never kept it this clean!"

President and Mrs. Hoover came to participate in the Jubilee. They arrived by train and were met by the Fords and the Edisons. They transferred to a small train and Edison's memory took hold of him. He began strolling up and down the aisle, calling out his wares of candy, apples, and sandwiches just as he had done as a boy.

Celebrities including Wilbur Wright and Marie Curie, who discovered radium, honored Edison by attending the Jubilee. Edison recreated the moment of the success of the light bulb by turning on the lights from the reconstructed laboratory.

President Hoover and others spoke, praising Edison for his work. Edison went to the podium next:

> In honoring me you are also honoring that vast army of thinkers and workers without whom my work would have gone for nothing. . . . This experience makes me realize as never before that Americans are sentimental. And this crowing event of Light's Golden Jubilee fills me with gratitude. As to Henry

Ford, words are inadequate to express my feelings. I can only say
to you, that in the fullest and richest meaning of the term—he
is my friend. Good night.

Edison's last series of experiments went outside the field
of electricity and into botany. Ford and Firestone had asked
him to help find a rubber substitute so they could keep manufacturing cars and tires. During World War I, natural rubber had been scarce. Much of the world's supply was under
British control, and Ford and Firestone wanted to find a way
to supply their needs from the United States.

Edison planted rubber trees in Florida and also tested
plants from his own yard and from around the country. In
spite of his eighty-one years, he performed numerous experiments. Mina Edison said, "Everything has turned to rubber
in our family. We talk rubber, think rubber, dream rubber.
Mr. Edison refuses to let us do anything else."

Rubber comes from a chemical known as latex that is
found in the sap of rubber trees. Starting with 14,000 plant
species, Edison tested each one for latex content. He narrowed
the possibilities down to 1,240 and from those, selected six
hundred plants that were easy to grow. Goldenrod was found
to have the highest latex content and as a weed, it was easi-
est to grow. Edison developed a type of goldenrod that was
twelve feet high. His last patent was for a process to extract
latex from goldenrod.

His process was never used. German and American chemists
were working on making synthetic rubber. In 1931, DuPont
patented a compound called neoprene, a synthetic rubber,
and tires were soon made up of one-half natural rubber and
one-half neoprene.

Edison in his Fort Myers, Florida garden in 1929, where he experimented with plants that could be used in the production of rubber. (*Courtesy of AP Images*)

Still, Edison's patent for the extraction process stood, and it was to be the last patent he filed. Throughout his career, he filed for and received 1,093 patents, more than anyone else before or since.

nine
Path of
Progress

Throughout his career, Edison had been snubbed by many scientists who viewed him as only an inventor, creating commercial products without great scientific value. With time, though, that attitude changed, and Edison finally found acceptance from his peers.

In 1911, he had been nominated for membership in the prestigious National Academy of Sciences. He only received three votes, so he was not chosen as a member. The then president of the Carnegie Institute, physicist Robert Woodward, said:

> the prejudice against men who follow the profession of engineering is especially strong unless they have been redeemed by that modicum of classical learning . . . We ought in our day to be entirely free of the notions which determine a man's standing by means of his early education or lack of it, and seek to measure men by their actual accomplishments.

As the years went on and Edison's accomplishments piled up, opinions of him changed among scientists. The president of the Massachusetts Institute of Technology (MIT), also a physicist, spoke about "Mr. Edison's Service for Science" in 1915. He called Edison "[one] who more than anyone else in this country has taught men to see something of what science can do."

In 1923, he became the first recipient of the Edison Medal awarded by the American Institute of Electrical Engineers, and received a gold medal from the Society of Arts and Sciences. In 1927, Edison was chosen to be a member of the National Academy of Sciences. A year later, Congress awarded Edison the Congressional Gold Medal with the inscription, "He illuminated the Path of Progress by His Inventions."

Edison's mother had taught him a love of books and he continued to enjoy reading. Even at the end of his life, he plowed through books. Mina said of him, "At home he always sits under the brilliant lamp reading as though he would devour all the books that were ever printed. He reads two or three lines at a time. Most of us read words—he reads whole sentences—more than whole sentences if they happen to be short ones. I have never seen anyone concentrate as he does."

His reading favorites ranged from Shakespeare and Victor Hugo to detective stories. In the 1890s, one author had used Edison as a model for his hero in a series of books. This character, Tom Swift, was created by Edward Stratemeyer, and he also wrote one of the Thomas Edison Jr. books that were popular at the time.

In his later years, Edison's hearing had worsened to the point of deafness. Mina often repeated to him what others

Edison operating a telegraph key on his eighty-third birthday. *(Courtesy of Mary Evans Picture Library/Alamy)*

had said. The tone of her voice and her slow speech allowed him to understand her.

Edison's poor diet and health habits had caught up with him. He suffered from diabetes and kidney problems. He had to remain in a wheelchair, but still worked in his laboratory.

The country was in the Great Depression in 1931. Edison, however as always, was optimistic. His health didn't allow him to attend a lighting convention, but he sent a message:

> My message to you is to be courageous. I have lived a long time. I have seen history repeat itself again and again. I have seen many 'depressions' in business. Always America has come

out strong and more prosperous. Be as brave as your fathers
were before you. Have faith—go forward.

Edison's health declined. He had been weak at the Light's
Golden Jubilee and his health continued to worsen. When he
left Florida that summer, he knew he would not return.

By the first part of October, Mina knew her husband was
fading. On October 14, Edison slipped into a coma. Early
on October 18, 1931, Thomas Edison died. He was eighty-
four years old. Reporters, who had been on a twenty-four
hour watch near Glenmont, sent the news around the country and the world.

His body was placed in an open casket in the library at
the West Orange laboratory. Thousands came to view the
inventor's body over the next two days. President Hoover
was in a conference with the premier of France and couldn't
attend the funeral, but Mrs. Hoover came. She joined the
Edison family, the Fords, and the Firestones as they listened
to the memorial service. The services were broadcast by the
National Broadcasting Service and included his favorite song,
"I'll Take You Home Again Kathleen." His body was laid
to rest at Rosedale Cemetery where New Jersey state police
kept a forty-eight hour vigil in his honor.

President Hoover requested a fitting tribute for the inventor. He asked that the country observe one minute of darkness and silence in Edison's honor. At 10 p.m. on October 21,
1931, lights all over the country were turned out, including
the torch on the Statue of Liberty.

Edison left a mixed family legacy. He was the father of
six children by two wives. His oldest three children often
resented the lack of their father's attention and presence.

Marion married a German army officer and lived in Germany through World War I, but she then divorced and returned to the United States in 1921. Thomas A. Edison Jr. was a big disappointment to his father. He occasionally used the Edison name to get money from people for all kinds of quack medicines and inventions. "I never could get him to go to school or work in the Laboratory. He is therefore absolutely illiterate scientifically and otherwise," expressed his father. William served in the military during the Spanish-American War and

Thomas Edison Jr. *(Library of Congress)*

during World War I. He later failed at farming. When his wife requested help and money, Edison said, "I see no reason whatever why I should support my son. He has done me no honor and has brought the blush of shame to my cheeks many times."

Edison was closer to his second set of three children, as he was around more during their childhoods. Madeleine attended Bryn Mawr College for two years, but disappointed her parents by marrying an actor. Four sons were born to the couple, the only grandchildren of Thomas Edison. At one time, Madeleine ran for Congress as a Republican like Thomas had been all his life. She worked passionately for the American Red Cross. After Mina died, Madeleine managed the Edison birthplace museum. In the 1950s, she served on the board of directors for Western Union, an echo of her father's beginnings. Charles ran his father's company, Thomas A. Edison, Inc. for thirty-two years, until it was sold. He also served in government as assistant secretary of the Navy and then acting secretary. He ran and won as a Democrat for governor of New Jersey. Theodore was the only one of the family to graduate from college. He received a degree in physics from MIT and worked in his father's company, eventually becoming laboratory director. After Thomas' death, Mina remarried and lived with her new husband at Glenmont as Mrs. Hughes. After his death, she retook the name of Mrs. Edison and remained at Glenmont until her death in 1947. In 1963, the remains of Thomas and Mina were moved from Rosedale Cemetery and interred at Glenmont.

Part of Edison's legacy can be seen in the fact that at his death, the lights could only be turned off for one minute.

Madeleine Edison *(Library of Congress)*

People had become too dependent on them to go longer than that. Edison's work prompted many to call him "Inventor of the World." His development of electrical components such as fuses, switches, and light bulbs put him in the forefront of the electrical industry.

His favorite invention, the phonograph, was the beginning of the multibillion dollar music industry. Edison also helped begin the movie industry by developing the movie camera and projector, and inventors after him meshed movement and sound into films, so that the industry grew into what we know today.

His impact can also be seen in experiments that he tried, but was unable to develop. Others stood on his shoulders and developed items that he had dreamed of. He was before his time on concrete houses, helicopters, alkaline batteries, vending machines, radio, and electronics.

Today, visitors can tour Edison's birthplace in Milan, Ohio, the Edison National Historic Site managed by the National Park Service in West Orange, New Jersey, and his winter home in Fort Myers, Florida. From January 26 to February 17, 2008, Fort Myers celebrated the 70th Annual Celebration of the Edison Festival of Light.

In 2004, the United States mint issued a commemorative Edison coin. The money raised by the sale of the coin was to be given to eight sites important to Edison. Efforts to preserve the Menlo Park Tower at the site of Edison's laboratory continued with funds from the coin sales.

In his birthplace of Ohio, people are campaigning for an Edison statue. The United States Capitol has an area known as Statuary Hall. Each state displays two statues of famous people. The current ones from Ohio are President Garfield and William Allen, a former governor, senator, and congressman. However, Allen voted against the Emancipation Proclamation, so in 2006, the state legislature voted to remove him and replace him with a statue of Edison. A plaque on it reads: Inventor of the Millennium.

Edison's success was supported by his curiosity about the world, his resilience when experiments failed, his optimism, and his hard work. His attitudes can be summed up by two of his most famous quotes: "There is no substitute for hard work;" "Genius is one percent inspiration and 99 percent perspiration."

Thomas Edison was born into a time that was moving faster than any before. He pushed progress forward at an even faster speed. He came into the Age of Steam and left in the Age of Electricity, an age he helped to usher in. Decades after his death, his work still illuminates the path of progress.

1847	Born in Milan, Ohio, on February 11.
1854	Family moves to Port Huron, Michigan.
1859	Gets a job at Grand Trunk Railroad; sets up a chemistry lab on the train.
1863 -67	Works as telegraph operator; begins experimenting with telegraph instruments.
1868	Files first patent for an automatic vote recorder.
1869	Patents several telegraph devices; moves to New York City, works for the Laws Gold Indicator Company.
1870	Opens a telegraph manufacturing shop.
1871	Marries Mary Stillwell.
1874	Invents the quadruplex telegraph, which transmits four messages simultaneously.
1876	Moves to Menlo Park, New Jersey; establishes industrial research laboratory.
1877	Invents the carbon transmitter; invents the phonograph.
1879	Invents carbon-filament lamp and generator for incandescent electric lighting; New Year's Eve demonstration of system held at Menlo Park.

1881	Opens offices in New York City; begins construction of the first permanent central power station, which opens September 1882.
1884	Wife, Mary, dies.
1886	Marries Mina Miller.
1887	Edison moves into a new laboratory in West Orange, New Jersey.
1892	The Thomson-Houston Company and Edison General Electric merge to form General Electric; attempts to develop method for processing low-grade iron and concrete.
1893	Demonstrates system for making and showing motion pictures.
1900	Begins work on storage battery for use in electric cars.
1915	Works with Naval Consulting Board to investigate new military technology.
1927	Tries to find a natural substitute for rubber that can be grown and processed quickly, eventually settling on goldenrod.
1929	Re-enacts the invention of the incandescent light at the Golden Jubilee celebration
1931	Dies in Llewellyn Park, New Jersey, on October 18.

sources

CHAPTER ONE: Young Businessman

p. 13, "he has always been . . ." Paul Israel, *Edison: A Life of Invention* (New York: John Wiley & Sons, Inc., 1998), 3.

p. 15, "I waited around for him . . ." Gene Adair, *Thomas Alva Edison: Inventing the Electric Age* (New York: Oxford University Press, 1996), 15.

p. 18, "My mother taught me . . ." Clair Price-Groff, *Thomas Alva Edison: Inventor and Entrepreneur* (New York: Franklin Watts, 2003), 10.

p. 19, "The world is my country . . ." Adair, *Thomas Alva Edison: Inventing the Electric Age,* 20.

p. 20, "My father had a set . . ." Israel, *Edison: A Life of Invention,* 8.

p. 20, "under all discouragements . . ." Price-Groff, *Thomas Alva Edison: Inventor and Entrepreneur,* 16.

p. 21, "hoeing corn in a hot . . ." Israel, *Edison: A Life of Invention,* 14.

p. 21, "I felt something in my . . ." Adair, *Thomas Alva Edison: Inventing the Electric Age,* 20.

p. 21, "I have not heard . . ." Ibid.

CHAPTER TWO: First-Class Telegrapher

p. 30, "Beethoven's greatest symphonies are . . ." Vincent Buranelli, *Thomas Alva Edison* (Englewood Cliffs, New Jersey: Silver Burdett Press, 1989), 21.

p. 31, "Any damn fool . . ." Adair, *Thomas Alva Edison: Inventing the Electric Age,* 34.

p. 31, "first-class man," Buranelli, *Thomas Alva Edison,* 21.

p. 33, "You seem to be tired . . ." Thomas Alva Edison, *The Papers of Thomas A. Edison: The Making of An Inventor,* (Baltimore: The Johns Hopkins University Press, 1989),1: 637.

p. 35, "I am now twenty-one . . ." Adair, *Thomas Alva Edison: Inventing the Electric Age,* 37.

p. 37, "T. A. Edison has resigned . . ." Ibid., 40.

CHAPTER THREE: Beginning Inventor

p. 39, "Fix it! Fix it!" Adair, *Thomas Alva Edison: Inventing the Electric Age,* 42.

p. 40, "Pope, Edison & Co., Electrical Engineers . . ." Price-Groff, *Thomas Alva Edison: Inventor and Entrepreneur,* 30.

p. 40, "A leading feature . . ." Ibid.

p. 41, "Best Electric Printing Telegraph Instrument," Israel, *Edison: A Life of Invention,* 54.

p. 41-42, "a young man of the highest . . ." Price-Groff, *Thomas Alva Edison: Inventor and Entrepreneur,* 31.

p. 42, "professor of duplicity and quadruplicity," Israel, *Edison: A Life of Invention,* 55.

p. 43, "Don't do any hard work . . ." Edison, *The Papers of Thomas A. Edison: The Making of An Inventor,* 1: 173.

p. 43, "I may be home . . ." Ibid., 212.

p. 43, "boys," Thomas Alva Edison to William Symes Andrew, 4 August 1883, the Thomas A. Edison Papers Project, Rutgers University. http://edison.rutgers.edu/

NamesSearch/SingleDoc.php3?DocId=LBCD1260.

p. 43, "old man," Adair, *Thomas Alva Edison: Inventing the Electric Age*, 57.

p. 43, "Mr H says that some . . ." Ibid., 219.

p. 44, "Mr. Edison, I can . . ." Israel, *Edison: A Life of Invention,* 73.

p. 44, "My Wife Popsy Wopsy . . ." Edison, *The Papers of Thomas A. Edison: The Making of An Inventor,* 460.

p. 49, "a radiant force, somewhere . . ." Adair, *Thomas Alva Edison: Inventing the Electric Age,* 93.

CHAPTER FOUR: Sound Inventions

p. 51-52, "The walls are covered . . ." Israel, *Edison: A Life of Invention,* 121.

p. 52, "a minor invention every ten . . . Ibid., 120.

p. 53-54, "Edison and his numerous assistants . . ." L. J. Davis, *Fleet Fire: Thomas Edison and the Pioneers of the Electric Revolution* (New York: Arcade Publishing, 2003), 214.

p. 54,56, "Mr. Watson, come here . . ." Adair, *Thomas Alva Edison: Inventing the Electric Age,* 59.

p. 57, "I rigged up an instrument . . ." Ibid., 60.

p. 58, "The machine must talk," Ibid., 61.

p. 58, "Finished the phonograph . . ." Price-Groff, *Thomas Alva Edison: Inventor and Entrepreneur,* 57.

p. 58, "the machine inquired . . ." Israel, *Edison: A Life of Invention,* 145.

p. 59, "I understand it all . . ." Price-Groff, *Thomas Alva Edison: Inventor and Entrepreneur*, 59.

p. 60, "The exhibition continued . . ." Ibid., 60.

p. 61, "a historic find . . . " Jody Rosen, "Researcher find song recorded before Edison's phonograph,"

International Herald Tribune (London), March 27, 2008.

p. 61, "squawk," Ibid.

p. 61, "appropriating," Ibid.

p. 61, "What are the rights . . ." Ibid.

p. 63, "this is my baby . . ." Israel, *Edison: A Life of Invention,* 47.

p. 63, "I can hardly wait until I get there," Ibid., 161.

p. 63, "the experiment had shown . . ." Ibid., 162.

p. 64, "only once was I . . ." Ibid.

p. 64, "the time has come when . . ." Ibid., 164.

p. 65, "He ran from the instruments . . ." Ibid., 165.

CHAPTER FIVE: Light Inventions

p. 66-67, "[W]e were trying . . ." Israel, *Edison: A Life of Invention,* 165.

p. 69, "money enough not only . . ." Ibid., 174.

p. 69, "I have it now! With . . ." Ibid., 168.

p. 69-70, "I have the right principle . . ." Ibid., 177.

p. 74, "the machine that will give . . ." Ibid., 176.

p. 74, "we have now . . ." Ibid., 182.

p. 76, "The light given was clear . . ." Jill Jonnes, *Empires of Light: Edison, Tesla, Westinghouse and the Race to Electrify the World* (New York: Random House, 2003), 61.

p. 76-77, "the train jumped the track, . . . Adair, *Thomas Alva Edison: Inventing the Electric Age,* 86.

p. 79, "The Great Inventor's Triumph . . ." Ibid., 81.

p. 79, "Menlo Park [was] thronged . . ." Israel, *Edison: A Life of Invention,* 187.

p. 80, "Extra trains were run . . ." Ibid.

p. 80, "how you got . . ." Adair, *Thomas Alva Edison: Inventing the Electric Age,* 10.

CHAPTER SIX: Lighting Up Manhattan

p. 82, "I hope that . . ." Randall Stross, *The Wizard of Menlo Park* (New York: Crown Publishers, 2007), 132.

p. 84, "Before I got through . . ." Price-Groff, *Thomas Alva Edison: Inventor and Entrepreneur,* 79.

p. 86, "*le grande* Edison," Jonnes, *Empires of Light: Edison, Tesla, Westinghouse and the Race to Electrify the World,* 73.

p. 86, "*C'est grand, c'est magnifique!*" Ibid.

p. 87-88, "with a clear, sharp . . ." Ibid., 75

p. 88, "The Pearl Street Station was . . ." Ibid., 81.

p. 88, "We're up in the world now," Ibid., 77.

p. 90, "Mr. Edison's exhibition is . . ." Price-Groff, *Thomas Alva Edison: Inventor and Entrepreneur,* 84.

p. 90, "At first everything worked all right," Adair, *Thomas Alva Edison: Inventing the Electric Age,* 97.

p. 90, "men who have battered . . ." Stross, *The Wizard of Menlo Park,* 133.

p. 91, "the lighting . . .was eminently . . ." Israel, *Edison: A Life of Invention,* 206.

p. 91, "I have accomplished all I promised," Ibid., 207.

CHAPTER SEVEN: New Industries, New Ideas

p. 93, "Mother was never happier . . ." Israel, *Edison: A Life of Invention,* 231.

p. 94, "shaking with grief, weeping . . ." Ibid., 230.

p. 97, "the most important part . . ." Adair, *Thomas Alva Edison: Inventing the Electric Age,* 102.

p. 98, "I'm going to do . . ." Israel, *Edison: A Life of Invention,* 339.

p. 98, "Well, it's all gone, . . ." Carlson, *Thomas Edison for Kids: His life and Ideas,* 105.

p. 99, "the brave builder of so gigantic . . ." Israel, *Edison: A Life of Invention,* 371

p. 102, "do for the eye . . ." Ibid., 107.

p. 107, "the quality of a cement . . ." Israel, *Edison: A Life of Invention,* 404.

CHAPTER EIGHT: World War I and Beyond

p. 109, "How would you like . . ." Israel, *Edison: A Life of Invention,* 309.

p. 111, "There's a young fellow . . ." Carlson, *Thomas Edison for Kids: His life and Ideas,* 126.

p. 111, "Young man . . ." "Henry Ford and Thomas Edison—A Friendship of Giants," *Detroit News.*

p. 111-112, "That bang on the . . ." Ibid.

p. 112, "This fellow Ford is like . . ." Carlson, *Thomas Edison for Kids: His life and Ideas,* 126.

p. 112, "I guess I'll have . . ." Ibid., 116.

p. 113, "Results! Why, man, I have . . ." Ibid., 118.

p. 114, "Where's Mother? Get her . . ." Adair, *Thomas Alva Edison: Inventing the Electric Age,* 120.

p. 114, "I am sixty-seven; but . . ." Price-Groff, *Thomas Alva Edison: Inventor and Entrepreneur,* 107.

p. 117, "I was an undergraduate . . ." Buranelli, *Thomas Alva Edison,* 103.

p. 117-118, "the people who worked . . ." Israel, *Edison: A Life of Invention,* 455.

p. 118, "all work on this plant . . ." Ibid., 356.

p. 118, "his greatest joy was . . ." Carlson, *Thomas Edison for Kids: His life and Ideas,* 128.

p. 118, "[he] always used to say . . ." Israel, *Edison: A Life of Invention,* 423.

p. 119, "He is constantly busy . . ." Carlson, *Thomas Edison f or Kids: His life and Ideas,* 129.

p. 119, "Something went wrong with . . ." Stross, *The Wizard of Menlo Park,* 256.

p. 120, "Well, you've got it . . ." Buranelli, *Thomas Alva Edison,* 117.

p. 120-121, "In honoring me you . . ." Adair, *Thomas Alva Edison: Inventing the Electric Age,* 126.

p. 120, "This experience makes me . . ." Buranelli, *Thomas Alva Edison,* 119.

p. 121, "Everything has turned to rubber . . ." Carlson, *Thomas Edison for Kids: His life and Ideas,*133.

CHAPTER NINE: Path to Progress

p. 123, "the prejudice against men . . ." Israel, *Edison: A Life of Invention,* 468.

p. 124, "[one] who more than . . ." Ibid.

p. 124, "He illuminated the Path . . ." Buranelli, *Thomas Alva Edison,*122.

p. 124, "At home he always . . ." Ibid., 69.

p. 125-126, "My message to you . . ." Carlson, *Thomas Edison for Kids: His life and Ideas,* 136.

p. 127, "I never could get . . ." "Thomas Alva Edison, Jr.," National Park Service, http://www.nps.gov/edis/historyculture/Thomas-alva-edison-jr.htm.

p. 128, "I see no reason . . ." Ibid.

p. 131, "There is no substitute . . ." "Thomas A. Edison Quotes," BrainyQuote.com, http://www.brainyquote.com/quotes/authors/t/Thomas_a_edison.html.

p. 131, "Genius is one percent . . ." Ibid.

bibliography

Adair, Gene. *Thomas Alva Edison: Inventing the Electric Age.* New York: Oxford University Press, 1996.

Buranelli, Vincent. *Thomas Alva Edison.* Englewood Cliffs, New Jersey: Silver Burdett Press, 1989.

Carlson, Laurie. *Thomas Edison for Kids.* Chicago, Illinois: Chicago Review Press, Inc., 2006.

Davis, L. J. *Fleet Fire: Thomas Edison and the Pioneers of the Electric Revolution.* New York: Arcade Publishing, 2003.

Edison, Thomas Alva. "A Christmas Past: Vintage Holiday Films: 1901-1923." New York: Kino Video, 2001.

———. *The Papers of Thomas A. Edison: The Making of An Inventor.* Edited by Reese J. Jenkins. Vol. 1. Baltimore: Johns Hopkins University Press, 1989.

"Inventors of the World: Thomas Edison." Wynnewood, Pennsylvania: Schlessinger Media, 2001.

Israel, Paul. *Edison: A Life of Invention.* New York: John Wiley & Sons, Inc., 1998.

Jonnes, Jill. *Empires of Light: Edison, Tesla, Westinghouse and the Race to Electrify the World.* New York: Random House, 2003.

Price-Groff, Clair. *Thomas Alva Edison: Inventor and Entrepreneur.* New York: Franklin Watts, 2003.

Stross, Randall. *The Wizard of Menlo Park.* New York: Crown Publishers, 2007.

Web sites

The Edison National Historic Site:
http://www.nps.gov/edis/index.htm

Edisonian Museum:
http://www.edisonian.com/index.html

Thomas Edison Web Page:
http://www.thomasedison.com/

Edison Innovation Foundation:
http://thomasedison.org/

The Edison Papers at Rutgers, the State University of New Jersey:
http://edison.rutgers.edu/about.htm

Thomas Edison in Menlo Park:
http://www.jhalpin.com/metuchen/tae/taeindex.htm

index